Guides to Professional English

Series Editor:
Adrian Wallwork
Pisa, Italy

For further volumes:
http://www.springer.com/series/13345

Adrian Wallwork

Email and Commercial Correspondence

A Guide to Professional English

 Springer

Adrian Wallwork
Pisa
Italy

ISBN 978-1-4939-0634-5   ISBN 978-1-4939-0635-2 (eBook)
DOI 10.1007/978-1-4939-0635-2
Springer New York Heidelberg Dordrecht London

Library of Congress Control Number: 2014939610

Printed on acid-free paper

Springer is part of Springer Science+Business Media (www.springer.com)

Who is this book for?

If you write emails and letters as part of your work, then this book is for you - particularly if you are a non-native speaker of English. By applying the suggested guidelines, you will stand a much greater chance of getting the desired reply to your emails in the shortest time possible.

I hope that other trainers like myself in Business English will also find the book a source of useful ideas to pass on to students.

This book is NOT for academics. Instead, read chapters 1–6 of *English for Academic Correspondence and Socializing* (Springer Science), from which some of the subsections in this book are taken or adapted.

Is this a book of the rules of email or just a guide?

It is only a guide.

It is based on 20 years of my own personal emailing experiences plus courses that I have held over the last 10 years at IT companies and research centers. They are ideas that my clients have found useful and which have significantly improved their relationships with the recipients of their emails.

I suggest you try out the various strategies outlined in this book: if they work for you, great. If they don't, then try another strategy. There is no one way that is guaranteed to be 100% effective.

What will I learn from this book?

Some of the key guidelines that you will learn from this book are:

- Write meaningful subject lines - otherwise recipients may not even open your mail.
- Always put the most important point in the first line - otherwise the reader may not read it.
- Never translate typical phrases literally - learn equivalent phrases.

- Be concise and only mention what is truly relevant. Write the minimum amount possible - you will also make fewer mistakes!

- Be a little too formal than too informal - you don't want to offend anyone.

- If you have two long important things to say, say them in separate emails.

- Give clear instructions and reasonable deadlines.

- Put in bold the most important thing you want recipients to read - then at least if they read the bold part and nothing else, you will still obtain what you wanted.

- If you need people to cooperate with you, you have to think of the benefits for them of cooperating with you.

- Empathize with your recipient's busy workload.

- Always be polite - and remember if there is a minimal chance that your reader will misinterpret or be offended then you can be sure he / she will. Never adopt an angry or super-authoritative tone. Saying 'please' is not enough.

How is the book organized?

Chapters 1–4 tell you how to organize an effective email from the subject line, initial salutation, body of the text, and final salutation.

Chapters 5–7 suggest ways for making requests and chasing people who have failed to do what you requested, and Chapter 8 deals with commenting on other people's work.

Chapters 9–12 focus on stylistic elements of writing emails: punctuation, formality as well as using a soft indirect style.

Chapter 12 deals with attachments.

Chapter 13 outlines the main differences between an email and a business letter.

Chapter 14 suggests ways to organize your emails and letters, avoid ambiguity, and make fewer mistakes from an English language point of view.

Chapter 15 is for fun only and lists typical abbreviations, acronyms and smileys.

The final two chapters list useful phrases for all types of emails (Chapter 16) and commercial letters and emails (Chapter 17).

How should I use the table of contents?

The table of contents lists each subsection contained within a chapter. You can use the titles of these subsections not only to find what you want but also as a summary for each chapter.

Other books in this series

There are currently five other books in this *Guides to Professional English* series.

CVs, Resumes, and LinkedIn
http://www.springer.com/978-1-4939-0646-8/

User Guides Manuals, and Technical Writing
http://www.springer.com/978-1-4939-0640-6/

Meetings, Negotiations, and Socializing
http://www.springer.com/978-1-4939-0631-4/

Presentations, Demos, and Training Sessions
http://www.springer.com/978-1-4939-0643-7/

Telephone and Helpdesk Skills
http://www.springer.com/978-1-4939-0637-6/

All the above books are intended for people working in industry rather than academia. The only exception is *CVs, Resumes, Cover Letters and LinkedIn*, which is aimed at both people in industry and academia.

There is also a parallel series of books covering similar skills for those in academia:

English for Presentations at International Conferences
http://www.springer.com/978-1-4419-6590-5/

English for Writing Research Papers
http://www.springer.com/978-1-4419-7921-6/

English for Academic Correspondence and Socializing
http://www.springer.com/978-1-4419-9400-4/

English for Research: Usage, Style, and Grammar
http://www.springer.com/978-1-4614-1592-3/

INTRODUCTION FOR THE TEACHER / TRAINER

Teaching Business English

I had two main targets when writing this book:

- non-native speakers (business, sales technical)
- Business English teachers and trainers

My teaching career initially started in general English but I soon moved into Business English, which I found was much more focused and where I could quickly see real results. The strategies I teach are almost totally language-independent, and many of my 'students' follow my guidelines when writing and presenting in their own language. I am sure you will have found the same in your lessons too.

Typically, my lessons cover how to:

1. write emails

2. make presentations and demos

3. participate in meetings

4. make phone calls

5. socialize

This book is a personal collection of ideas picked up since the advent of email. It is not intended as a course book, there are plenty of these already. It is more like a reference manual.

I also teach academics how to present their work. In fact, some of the chapters in this book are based on chapters from *English for Academic Correspondence and Socializing* (Springer).

How to teach email writing

I suggest you adopt the following strategy.

First, get your students to send you collections of around 10 emails that they typically write (and receive). Before they send them to you they should obviously delete any confidential information regarding their company. Make sure they also include subject lines.

Read through the collections of emails and convert them all into one Microsoft Word (or equivalent) document. Having the email examples in Word will enable you to revise them using Track Changes during class (i.e. with your laptop connected to a bigger screen so that students can see clearly).

In your first lesson on email, have a general discussion on:

- how they choose their email address and what effect they think this has on recipients (Chapter 1)

- how much time your students spend on writing and dealing with emails

- whether they write emails from scratch or use Google Translate

- how much effort they make to ensure their emails are well constructed and in perfect English; and compare this level of effort with the effort they make when writing in their own language

- how formal / informal (Chapter 11) their email correspondence is and how it differs from a business letter (Chapter 13)

- what they think are the qualities of a good email

- what useful phrases (Chapters 16 and 17) they know, and whether they have made their own personal collections of such phrases

In the next lesson, focus on subject lines (read Chapter 2 in this book while preparing for your lesson). Collect 10-20 subject lines, then in the lesson get students to decide which ones are effective and why. They can then improve the ones that they think are less effective.

Then move on to salutations (Chapters 3 and 4), using the same strategy as with subject lines.

Now you will need to turn to the main body of the email. Most emails consist of either making requests (Chapter 5) or replying to them (Chapter 6).

When your students have mastered the basics of email writing (Chapters 1–6, 10–13), with more advanced students you can then deal with difficult emails: chasing people (Chapter 7) and criticizing (Chapters 8 and 9).

To discuss how to organize an email, use Chapter 14, which also covers avoiding ambiguity and English language mistakes.

You can have some fun by showing them how English can be used in weird and wonderful ways - see Chapter 15 (on text messaging style and smileys).

When you are preparing each lesson, make sure you find examples from your students' own emails, i.e. the ones that you collected in the first stage of preparing your email course. You can also create a variety of exercises using their emails, for example on grammar, useful phrases, and level of formality.

Finally, get involved with the company / companies where you teach. You will find your work much more satisfying!

Contents

1.1 Choose a suitable email address

If you email someone for the first time (i.e. a new contact), he / she will open your email on the basis of:

1. your email address

2. how your name appears in their inbox

3. your subject line (see Chapter 2)

Your email address reflects your level of professionalism. Avoid any of the following types of address:

lordofdarkness@yahoo.com (name of favorite rock band, movie etc)

andrew1999@hotmail.com (first name + number / date of birth)

verwhite@gmail.com (merge of first name and second name, i.e. Veronica White)

Instead, clearly differentiate your first name from your last name. Here are my personal and work addresses:

adrian.wallwork@gmail.com

adrian.wallwork@e4ac.com

They look professional and no one is going to get a negative impression from them. Also they make it easy for recipients to find your address within their email system. On the other hand, if I wrote my address as follows (i.e. not given name + family name), it might be much more difficult for my recipient to locate me:

adrianwallwork@gmail.com

awallwork@gmail.com

wallworka@gmail.com

A. Wallwork, *Email and Commercial Correspondence,*
Guides to Professional English, DOI 10.1007/978-1-4939-0635-2_1,
© Springer Science+Business Media New York 2014

1.2 Check how your name will appear in your recipient's inbox

If your recipient, like me, uses Gmail, then the first time your recipient receives an email from you, the first two words of how your choose your email name will appear, in my case *Adrian Wallwork*. In future emails, it will appear with just the first word.

This means that if you have chosen the name *Prof. Adrian Wallwork*, then in future emails your recipient will just see the word *Prof.* Consequently, when choosing your name as it will appear in your recipient's inbox, do not add any titles.

If you are a company, and you have an address such as *info@e4ac.com* (where e4ac is the name of the company, in this case, the name of my company), then be aware that recipients will just see *info* in their inbox and will have no idea what your company is called.

1.3 Avoid a spam-friendly email address

Before choosing an email address and name, look through your spam. Note the kinds of addresses that are found there. Make sure that yours does not look similar in type. Not having a suitable email address will increase the likelihood that your email will end up in your recipient's spam (see 2.8).

2 SUBJECT LINES

2.1 Compose the subject line from the recipient's perspective

A good subject line should be written from the recipient's perspective

- be pertinent to the recipient (not just to you)

- encourage the recipient to open the mail itself

- indicate to the recipient whether he / she needs to open it immediately or later

- easily searchable / retrievable by the recipient

- short

- give a very clear indication of the actual message

Also, remember that the email you have written may then become part of a long chain, possibly with multiple recipients. If possible, choose a subject line that will not need to be changed (because it is not sufficiently pertinent) at some point later in the chain.

2.2 Combine your subject line with the preview pane

Most email systems display not only the subject line but also make the first few words visible too. It may be useful to use the first words as a means to encourage the recipient to open your email straight away.

If you adopt this tactic, then it is a good idea to keep your subject line as short as possible. If you can include any key words in the first few words, that too will have a positive influence on the recipient.

A. Wallwork, *Email and Commercial Correspondence,*
Guides to Professional English, DOI 10.1007/978-1-4939-0635-2_2,
© Springer Science+Business Media New York 2014

2.3   Include name of a mutual third party in the subject line (and in cc) in a mail to a new contact

Imagine you are writing to a totally new contact and that this contact knows someone who knows you. In such cases, it is not a bad idea to put the name of this common acquaintance in the subject line. This alerts the recipient that this is not a spam message. For example, let's imagine you met Jo Bloggs at a trade fair and you subsequently became connected on LinkedIn. Bloggs recommended that you write to a colleague of his, Andrea Wilkes (who does not know you), for a possible interview at Wilkes' company. Your subject line for your email to Andrea Wilkes could be:

Jo Bloggs. Request for interview for position as software analyst

From the subject line, Andrea knows that the message is not spam, and by seeing the name of someone who is familiar to her (Jo Bloggs) she will be motivated to open the email. Andrea will be further motivated to open your email if she sees that Jo Bloggs is in cc.

2.4   Make it clear if your mail only requires very limited effort on the part of the recipient

Many people don't open an email on the basis that it will require time on their part to answer the mail. However, if you can make it clear in the subject line that in fact no effort will be required other than reading a couple of lines of text and then giving a simple yes / no answer, then your recipient is more likely to open the mail and respond. For example:

Two minutes of your time: Could you check the attached figures? Thanks.

Quick favor: Can you update the sales figures on the attached Excel file?

2.5   Be specific, never vague

A vague subject title such as *Meeting time changed* may annoy many recipients. They want to know which meeting, and when the new time is. Both these details could easily be contained in the subject line.

Project C Kick Off meeting new time 10.30, Tuesday 5 September

This means that a week later when perhaps your recipients have forgotten the revised time of the meeting, they can simply scan their inbox, without actually having to open any mails.

2.6 Consider using a two-part subject line

You can split your subject line into two parts. The first part contains the context, the second part the details about this context. Here are two examples:

XYZ meeting: new time 10.00

Annual review: 10 Nov deadline approaching

2.7 Ensure that your subject line is not spam friendly

Very generic subject lines often cause an email to end up in the spam. This is particularly true if they are combined with a generic salutation at the beginning of the email. Below is an example of a mail from a teacher in Iran that was intended specifically for me, but in fact was marked as spam by Gmail.

Subject line: kindly appreciate your help

Dear sir,

I'm an English teacher and for a writing course I have been asked to teach your precious book on writing papers (English for Writing Papers). I deeply appreciate your kindness if you could guide how I can teach this book.

The subject line above is typical of thousands of spam mails that attempt to get people to make illegal bank transfers. Moreover, native English speakers do not begin an email with *Dear Sir*. Interestingly, the email address of the sender was an academic one (@alumni.ut.ac.ir), which highlights that Gmail focuses on the text of the mail rather than the address of the sender.

A more effective email that would not have ended up in my trash, would be:

Subject line: English for Writing Papers – help with methodology

Dear Adrian Wallwork,

I'm an English teacher and ...

The revised version above has

- a clear specific subject line
- a salutation that includes my name

Both these factors are crucial in ensuring your email does not go in the recipient's spam.

3 INITIAL SALUTATIONS

3.1 Avoid gender titles (e.g. Mr, Mrs) in first email to new contact

It may be difficult to establish someone's gender from their first name. In fact, what perhaps look like female names, may be male names, and vice versa. For example, the Italian names Andrea, Mattia and Nicola; the Russian names Ilya, Nikita, and Foma; and the Finnish names Esa, Pekka, Mika and Jukka are all male names. The Japanese names Eriko, Yasuko, Aiko, Sachiko, Michiko, Kanako may look like male names to Western eyes, but are in fact female. Likewise, Kenta, Kota, Yuta are all male names in Japanese.

In addition, many English first names seem to have no clear indication of the sex e.g. Saxon, Adair, Chandler, Chelsea. And some English names can be for both men and women e.g. Jo, Sam, Chris, and Lesley.

So be careful how you use the following titles:

Mr – man (not known if married)

Ms – woman (marriage status irrelevant)

Mrs – married woman

Miss – unmarried woman

If you really consider it essential to use a title, then the safest option is to use Mr or Ms, for males and females respectively.

In some cases it may not be clear to you which is the person's first and last name, e.g. Stewart James. In this particular case, it is useful to remember that Anglos put their first name first, so Stewart will be the first name. However, this is not true of all Europeans. Some Italians, for example, put their surname first (e.g. Ferrari Luigi) and others may have a surname that looks like a first name e.g. Marco Martina. In the far east, it is usual to put the last name first, e.g. Tao Pei Lin (Tao is the surname, Pei Lin is the first name). The best solution is always to write both names, e.g. Dear Stewart James, then there can be no mistake. Similarly, avoid Mr, Mrs, Miss and Ms – they are not frequently used in emails. By not using them you avoid choosing the wrong one.

A. Wallwork, *Email and Commercial Correspondence*,
Guides to Professional English, DOI 10.1007/978-1-4939-0635-2_3,
© Springer Science+Business Media New York 2014

3.1 Avoid gender titles (e.g. Mr, Mrs) in first email to new contact (cont.)

If your own name is ambiguous, it is a good idea in first mails to sign yourself in a way that is clear what sex you are, e.g. Best regards, Andrea Cavalieri (Mr).

A good general rule when replying to someone for the first time is to:

- address them using exactly the same name (both first and last name) that they use in their signature

- precede this name with an appropriate title

- adopt their style and tone. If you are making the first contact then it is safer to be formal in order to be sure not to offend anyone.

Then as the relationship develops, you can become less (or more) formal as appropriate. In any case, always take into account the reader's customs and culture, remembering that some cultures are much more formal than others.

3.2 Spell the recipient's name correctly

Make sure your recipient's name is spelt correctly. Think how you feel when you see your own name misspelled.

Some names include accents. Look at the other's person's signature and cut and paste it into the beginning of your email – that way you will not make any mistakes either in spelling or use of accents (e.g. è, ö, ñ).

Although their name may contain an accent, they may have decided to abandon accents in emails – so check to see if they use an accent or not.

3.3 Use 'Dear + first name + second name' only in the first contact

If you were writing to me (the author of this book) for the first time you would write:

Dear Adrian Wallwork

But you would not use this formula for every subsequent email that you write, i.e. you would not write *Dear Adrian Wallwork* in your next email.

Depending on the level of formality that you wish to maintain and also depending on how I sign myself in my reply to you, then you would probably write:

Dear Adrian – if I signed myself simply *Adrian,* or if the tone of my email is very friendly and you believe that we are on a similar level in terms of company hierarchy

Dear Mr Wallwork – if I signed myself *Adrian Wallwork,* or if the tone of my email is neutral or quite formal and you believe that you are on a lower level in terms of company hierarchy

3.4 Be careful of punctuation

You can punctuate the salutation in three ways:

Dear Adrian Wallwork,

Dear Adrian Wallwork:

Dear Adrian Wallwork

i.e. with a comma, with a colon, or with nothing. Whichever system you use, you always begin the first line of the body of the email with an initial capital letter. Example:

Dear Adrian,

Thank you for getting in touch with me …

3.5 Choose a specific job title when addressing an email to someone whose name you do not know

For important emails it is always best to find out the name of the person to address. This maximizes the chances of your email (i) reaching the right person, (ii) being opened, and (iii) being responded to.

However, on many occasions the exact name of the person is not important, for example, when you are contacting a helpdesk. In such cases, the simplest solution is to have no salutation at all, or simply to use *Hi*. Some people like to use the expression *To whom it may concern* but this expression is really no more useful than having no salutation.

Alternatively, you can write something more specific, such as:

Dear Sales Manager

4 INTRODUCTIONS AND FINAL SALUTATIONS

4.1 Explain where you got your contact's details from

When writing to a new contact you will probably want to inform the recipient how you got their email address. Examples:

I found your details on LinkedIn, we have a connection in common – Joe Amos.

I was given your CV by a colleague, Tao Pei Lin. As you may know, we are looking for a sales manager in our office in Beijing and I would like to discuss this position with you further.

Your name was given to me by …

A. Wallwork, *Email and Commercial Correspondence,*
Guides to Professional English, DOI 10.1007/978-1-4939-0635-2_4,
© Springer Science+Business Media New York 2014

4.2   Introduce yourself to a new contact

When you are writing to a new contact, either within your own company or externally, your first line can be used to make an initial introduction

> Just a couple of lines to introduce myself. My name is Kristina Kurtis and I am responsible for marketing for ABC globally. I'm based in Athens and I ...

> My name is Nirupa Kudahettige and I'll be your contact for any issues related to quality control.

> Just a quick introduction: my name is Mohammed Abdelwahab, I head up business development here in Oman.

> Lora is away until Monday, and I am covering her work. I have been asked to ...

Clearly, there is no real need to say your name as this will be evident immediately from the recipient's inbox, however it is general practice to do so.

Note the use of the following prepositions:

> FOR a company – I work for Google.

> IN a city / town / office – I work in the New York office.

> UNDER a person (i.e. your boss / supervisor) – I work under Mike Jackson.

> IN a team / department – I work in the development team in the R&D department.

> ON a project – I am working on the XYZ project.

> be responsible FOR / in charge OF – I am responsible for client relations. I work under Mike Jackson who is in charge of marketing

4.3 Give details about who you are and what you are requesting

Below are some examples of first emails in which the sender introduces themselves and then gives details regarding the reason for the email.

This email is to someone external to the sender's company.

Hi Zach

You may remember that we worked together back in 2013 when you were with Top Recruitment Solutions for the placement of a trainee sales clerk for our accounts department.

I am contacting you because we need to recruit some technical staff (business development and analysts) for our office in Riyahd. I know that this is not your specific field but I was hoping you might be able to put me in contact with the right person.

Looking on LinkedIn I see that you now work for Saudi Recruitment International, so I was wondering whether there might be a chance of setting up a collaboration with your new agency.

Perhaps we could set up a Skype call to discuss this.

Best regards

This email is to someone within the sender's company.

Dear Greg

My name is Mercy Boatemaa. I recently started working on a variety of corporate strategy initiatives. I have been asked to kick off the quality systems project. At the moment, I am just collecting information on what is currently in place and what the users of the system would like it to be able to do. Given your role, I would love to pick your brains on this process. Would you be free to discuss requirements, current workflows, etc. in further detail later this week?

I'm looking forward to working together.

4.4   Introducing someone to a third party

Sometimes you may need to put two people in contact with each other. You will probably do this by addressing the same mail to both recipients, one of whom maybe in copy (cc). Ensure that you explain why you are putting the two parties in contact.

> Dear Henri
>
> This is to introduce Ahmed, who is in cc, our new employment solicitor in Dubai. From now on Ahmed will be dealing with ... and you should contact him with any queries.

4.5   Reminding a contact who you are

You can announce your name and where you met.

> My name is Heidi Muller and you may remember that we met in the VIP lounge at Heathrow Airport last week. I asked you the question about X. Well, I was wondering ...

Or without announcing your name you can simply jog their memory.

> Thanks for the advice you gave me at dinner last night. With regard to what you said about X, do you happen to have any reports on ...

For someone you collaborated with several years ago, you can remind them of the context.

> You may remember that we worked together back in 2013 when you were with Dean Solutions in the Finance & Accounting department.

4.6 Begin with a greeting + recipient's name

If you begin an email simply with 'hi' or 'good morning' or with no greeting at all this will not help the recipient know if the message really was intended for them. Given that your recipient will be able to see the beginning of your email without actually opening the email, if they do not see their name they may think that either that the message is not for them, or that it is spam, and thus they may delete it without reading it.

A greeting provides a friendly opening, in the same way as saying 'hello' on the phone. A greeting only requires a couple of words, and on the recipient's part will take less than a second to read so you will not be wasting their time. Example:

Hi Adrian,

How are things?

However, if you exchange messages regularly with someone and that person does not make use of greetings, then you can drop these greetings too.

4.7 Indicate which of the multiple recipients actually needs to read the mail

If someone is on a mailing list they may receive hundreds of emails that are not specifically for them. It is thus good practice to begin your email by saying who exactly the email is for and why they should read it, then those who may not be interested can stop reading.

4.8 If in doubt how to end your email, use *Best regards*

There are many ways of ending an email in English, but the simplest is *Best regards*. You can use this with practically anyone.

Best regards is often preceded with another standard phrase, for example *Thank you in advance*, or *I look forward to hearing from you*. For more standard phrases see Chapter 16.

Note the punctuation. Each sentence ends with a full stop, apart from the final salutation (Best regards) where you can put either a comma (,) or no punctuation.

I look forward to hearing from you.	Thanks in advance.
Best regards,	Best regards
Adrian Wallwork	Adrian Wallwork

4.9   Don't use a sequence of standard phrases in your final salutation

When writing emails in your own language you may be accustomed to using a sequence of standard phrases at the end of your emails. This is not common practice in English.

Imagine you need to ask someone for a favor. When writing to North Americans, British people, Australians etc, normally two phrases would be enough in your final salutation. For example:

> Thank you very much in advance.
>
> Best regards
>
> Syed Haque

The above email is polite and quick to read. The following email contains too many salutations and is also rather too formal.

> I would like to take this opportunity to express my sincere appreciation of any help you may be able to give me.
>
> I thank you in advance.
>
> I remain most respectfully yours,
>
> Syed Haque

Bear in mind that many business people receive up to 100 emails a day, thus they do not have time to read such a long series of salutations.

4.10 Ensure your signature contains everything that your recipient may need to know

What you include in your signature has some effect on the recipient's perception of who you are and what you do. It is generally a good idea to include most or all of the following.

- Your name
- Your position
- Your company
- Your department / division / branch (both in English and your mother tongue)
- Your phone number
- The switchboard phone number of your department

Make sure your address is spelt correctly and that you have correctly translated the name of your department.

4.11 Avoid PSs and anything under your signature

When recipients see your salutation (e.g. *Best regards*) or name it is a signal for them to stop reading. If you write a PS (i.e. a phrase that is detached from the main body of the mail and which appears under your name) or anything under your signature, there is a very good chance it will not be seen / read.

5 MAKING REQUESTS

5.1 Decide whether it might be better just to make one request rather than several

If you have one particular important thing to ask, only ask that one thing.

If you have only one request in your email, the recipient will have fewer options – he or she will either ignore your email, or will reply with a response to your request. The fewer options you give your recipient, the more likely you will achieve what you want.

Do not add other requests within the same email. Generally speaking, when we receive several requests within the same email, we tend to respond to the request or requests that is / are easiest to deal with, and ignore the others.

5.2 Lay out your request clearly and give recipients all the information they need to carry out the request

To ensure that your recipients follow your requests, you need to motivate them to do so. You can do this by providing:

- clear instructions of what you wish them to do

- reasons why your request is relevant to them

Below is an email telling team leaders about the annual review process in their company.

Please find attached the annual team reviews spreadsheet for this year.

The spreadsheet contains:

- Instructions on how to use the form and the review process this year.

- Attributes to rate your team members.

- A team summary sheet.

A. Wallwork, *Email and Commercial Correspondence,*
Guides to Professional English, DOI 10.1007/978-1-4939-0635-2_5,
© Springer Science+Business Media New York 2014

- A sheet to complete with your review comments about the team member.
- A marker sheet to insert self-assessment comments.

Please read the instructions in the spreadsheet before completing your team review.

Note: Please return the completed spreadsheet by Wednesday December 5 at the latest.

If you have any questions, please e-mail: annualreviews@thecompany.com

Thank you for your cooperation. Your contribution will help your team improve its performance in the coming year.

In the above email note:

- the clear structure and layout (including the use of bullets)
- the clear instructions
- the use of bold to highlight the importance of one particular instruction
- the provision of a contact address for queries
- the final line which is designed to thank the recipient and help them understand the importance of completing the task

5.3 Avoid blocks of text and don't force your reader to make sense of everything

In the case below the sender is requesting some product information. However, she is seriously jeopardizing the chances of receiving an answer. In fact, she has written one long block of text containing a considerable amount of information that is of little or no interest to the recipient. The recipient only needs to know the exact details of the sender's request.

Hi

I'm Maria Masqueredo and I work for ABC. I am currently working on a project that entails the use of shape memory alloy tubes and a colleague of mine referred me to your website where I found a few examples that might satisfy my requirements. Essentially, I need shape memory alloy tubes (not superelastic alloys). The transformation temperature is not a critical parameter (Af = 70 C or more would be adequate). What is really important is that the ratio between the internal diameter, di, and the external diameter, de, must be near the value of 0.7–0.8. The external diameter can be 1.5 mm or more (not exceeding 12 mm). Do you have any product able to satisfy my constraints? Can you send me an estimate for 5 m of your products? By the way I found a mistake in one your product descriptions, under 'steel tubes' I think it should say 'alloy' rather than 'allay'.

Thank you in advance for any help you may be able to give me.

Best regards

A better version would be:

Hi

Do you have a shape memory alloy tube with the following characteristics?

1. transformation temperature of Af = 70 C or more

2. ratio between the internal diameter and the external diameter must be 0.7–0.8

3. external diameter in a range from 1.5 mm to 12 mm

If so, please could send me an estimate for a 5 m tube.

Thanks in advance.

Maria Masqueredo

In the original example above Maria has not thought about the recipient. She has simply written down her thoughts as they came into her head, thus leaving it to the recipient to make sense of everything. If the recipient has the time to deal with the email he / she might answer it, but there is a good chance that he / she will leave it till later or simply delete it on the basis that it is not time-efficient or cost-effective to deal with it.

5.4 Make all your requests 100 % clear

If you ask multiple questions within the same email, you need to lay out and structure your email very carefully. If you don't do this, you are unlikely to get answers to all your questions, but probably only to those questions that your recipients can see the most quickly or which require the least effort on their part.

So make your requests absolutely clear.

Here is an email I received from the permissions department of a publisher. I wanted to know if I could use a short piece of text from one of their books in one of my own books. They replied as follows:

> Please let me know how many copies of the book are being printed, where they will be sold (what territories) and what is the term of license under section 4779.09 of the Revised Code for this book?

There are two problems with this request. First, there are three requests in one sentence. For recipients this is a problem, because they cannot quickly identify the requests when replying to them. Secondly, it includes the phrase 'term of license under section 4779.09 of the Revised Code'. This phrase was probably very clear for the sender (i.e. the publisher) because it relates to their field of business, but it meant nothing to me – it was too technical. My choices were (i) try and find out the meaning on the web, (ii) ask for clarification by writing another email, or (iii) just to ignore it completely and simply answer the other two questions.

Basically, most recipients will opt for what seems to be the easiest solution, which would be the third solution – ignore the request. So if you are making a request, ensure that you phrase it in such a way that your recipient will have no problem understanding it and will thus

- not need to ask for or look for clarification (and thus not waste further time)
- respond to your request, hopefully with information you wanted

As always, think in terms of your recipient and not of yourself.

A clearer version of the above email could be:

> Please could you kindly answer these three questions:
>
> 1. how many copies of the book are likely to be printed per year?
> 2. what territories will they be sold in?
> 3. what is the term of license for this book (i.e. when will the contract for the book expire)?

The revised version alerts the recipient that there are three requests to answer, and underlines this by using numbered bullets. The first question is also more precise (*per year*) and the third question now includes an explanation of the technical phrase (i.e. *term of license*) and has simply deleted the reference to the section of the Revised Code as being unnecessary.

Clearly, the revised version would take more time to write than the original version, but the benefit is that the writer is more likely to get replies to all three questions.

To ensure that all your requests get answered, it is generally wise to number them and keep them as short as possible.

When writing requests and instructions, particularly to multiple recipients, you also need to make it 100 % who is expected to carry out which tasks.

For example, what should the recipient understand from the following? (the use of bold is mine)

> ... clients are not happy about this. So **it is possible that the tests will have to be repeated** again.

> ... clients are not happy about this. So **it is necessary to repeat** the tests again.

> ... clients are not happy about this. So the tests **need to be repeated** again.

It would be much better to write:

> ... clients are not happy about this. So **if you don't mind I think you** need to repeat the tests again. // So I am afraid **you will have to** repeat the tests again.

The problem with impersonal phrases such as *it is possible / probable / necessary / mandatory* and with passive forms is that the subject of the action is not mentioned, so the recipient does not know if it is him / her who has to carry out the task or someone else.

5.5   For multiple requests, include a mini summary at the end of the email

Many recipients only read the email once. This means that by the end of the email they may have already forgotten any requests that were made at the beginning of the email. Thus they may respond to only the request / s that they remember or simply the ones that are easiest for them to deal with. This happens even if you have used bullets and used lots of white space to indicate a clear division between your requests.

The email below illustrates some techniques that may help you to increase your chances of getting a reply.

Dear Yohannes

I hope you had a good a holiday. I have three short requests that I hope you might be able to help me with.

REQUEST 1

Can we meet to discuss the client's specifications next week, preferably Wednesday morning. Basically I want to focus on blah blah blah blah blah blah blah blah blah blah blah blah blah blah blah blah blah blah blah blah blah blah blah blah blah blah blah blah blah blah blah blah.

REQUEST 2

Do you have a copy of the XYZ report? If so, could you get it me by tomorrow lunch time. The reason I need it is blah blah blah blah blah blah blah blah blah blah blah blah blah blah blah blah blah blah blah blah blah blah blah blah blah blah blah blah blah blah blah.

REQUEST 3

Has the ABC spreadsheet updated? Can you send the latest copy please.

Summary:

1. Meet next Weds morn – client specifications.

2. XYZ report by tomorrow lunch.

3. ABC spreadsheet.

I look forward to hearing from you.

The techniques are:

- precede each request with a number (Request 1, Request 2 etc) and put the word request in capital letters so it clearly stands out

- provide a summary of all the requests at the end; put the word 'summary' in bold

Generally speaking you would only need to use one of the two techniques, particularly if the email is reasonably short as in the example above. But if an email is long and requires scrolling by the recipient, then a summary at the end will certainly increase the chances of your recipient answering all your requests. The summary also helps the recipients as they can simply insert their answers under each point of the summary.

Obviously, there are other equally effective ways to achieve the same objective. For example:

Dear Yohannes

I hope you had a good a holiday. I have three short requests that I hope you might be able to help me with.

1. Meet next Weds morn – client specifications

Can we meet to discuss the client's specifications next week, preferably Wednesday morning. Basically I want to focus on blah blah blah blah blah blah blah blah blah blah blah blah blah blah blah blah blah blah blah blah blah blah blah blah blah blah blah blah blah blah blah blah.

2. XYZ report by tomorrow lunch

Do you have a copy of the XYZ report? If so, could you get it me by tomorrow lunch time. The reason I need it is blah blah blah blah blah blah blah blah blah blah blah blah blah blah blah blah blah blah blah blah blah blah blah blah blah blah blah blah blah blah blah blah.

3. ABC spreadsheet.

Has the ABC spreadsheet updated? Can you send the latest copy please.

Best regards

5.6   Give deadlines

You will increase your chances of people responding to your requests if you give them a specific deadline. This is much more effective than saying: *as soon as possible* or *at your earliest convenience*, as these two phrases given no idea of the urgency of the sender.

However, it pays to give recipients a reasonably short deadline and not too many options. The longer the deadline you give them the greater the chance that they will simply not remember to fulfill your request. Typical phrases you can use are:

I need it *within* the next two days.

He wants it *by* 11 tomorrow morning at the latest.

I don't actually need it *until* next week, Tuesday would be fine.

I need it some time *before* the end of next week.

Note how the words in italics are used in the context of deadlines.

within to mention a period of time, which is always indicated by a plural noun (hours, weeks, months).

by to indicate a specific moment in the future which is the end point of a period of time during which something must be done

until with negative verb (e.g. *I don't need x until y*) to mean 'not before'

before the same as *by*, but *by* can also mean *at*, whereas *before* can only mean 'at any point during a period of time'

If you are the receiver of a deadline or if you simply wish to establish your own deadline, then you can use similar phrases. For example, if someone writes to you saying *Could you revise the section as soon as possible*. You can say:

> I should be able to get the revisions back to you *by* the end of this month / *within* the next ten days.

> I am sorry but I won't be able to start work on it *until* Monday / *before* next week at the earliest.

5.7 Motivate the recipient to reply by empathizing with their situation or by paying them a compliment

Most recipients are more likely to meet your requests if you seem to show some understanding of their situation or if you appreciate their skills in some way. Here are some typical phrases that senders use to motivate their recipients to reply.

> I know that you are very busy but …

> Sorry to bother you but …

> I have heard that you have a mountain of work at the moment but …

> Any feedback you may have, would be very much appreciated.

> I have an urgent problem that requires your expertise.

> I really need your help to …

> I cannot sort this out by myself …

6.1 Apologizing for late reply

Typical excuses for not replying promptly are:

> Sorry but for some reason my system thought your email was spam.

> Sorry I was convinced I had already replied.

> Sorry but I have been out of the office all week.

If you don't feel it is necessary to have any excuse you can say:

> I apologize for not getting back to you sooner.

Avoid giving an excuse that is likely to irritate the recipient or make you seem inefficient, such as the following:

> I apologize greatly for the delay. I have had a week away skiing and did not put my out of office reply on. I did ask my colleagues to keep an eye on all incoming emails so I am sorry nobody got back to you.

When you are late in replying, you can apologize both at the beginning and the end of the email:

> Please accept my apologies, I was convinced that I had replied to you. Your best bet to get info on this is to ask Yuki directly – he is in the London office. Thanks and once again sorry for not getting back to you straight away.

If possible, state what you are doing to resolve the situation that the sender has informed you about.

> My sincere apologies ... I am still in the process of trying to find the information for you.

> I am genuinely very sorry about the delay on this Robert, I will get the documents to you as soon as I possibly can.

A. Wallwork, *Email and Commercial Correspondence,*
Guides to Professional English, DOI 10.1007/978-1-4939-0635-2_6,
© Springer Science+Business Media New York 2014

6.2 Consider inserting your answers within the body of the sender's email

There are basically two ways of replying to an email:

- write your reply under the sender's text
- insert your replies within the sender's text

Let's imagine that you are Raul, and you work in the Madrid office. Peter, who works in the NY office, sends you the email below.

Hi Raul

I hope all is well with you. I was wondering if you could do me a couple of favors. Attached are two documents. The first is a Request for Proposals that I would like you to read and hear your comments on. There is actually a 500 word limit and it is currently around 650, so if you could find any way to remove a few words that would be great. Also attached is a proposal for the Request for Funding – for some reason I can't find the email addresses of the relevant people in your office in Madrid, so could you possibly forward it to them? Thanks. Then finally, you mentioned last time we talked that you had a useful contact at a recruitment agency that you thought I should look up, do you think you could send me the name? Thanks very much and sorry to bother you with all this.

If we don't speak before, I hope you have a Happy Christmas!

Best regards

Peter

You could decide to write your reply under Peter's complete text as follows:

VERSION 1

Hi Pete

Good to hear from you. Yes, I am happy to read the RFP and I will try to reduce the word count. I have forwarded the request for funding proposal and I put you in cc. Please find below the name of the recruitment person I mentioned: Romeo Henandez (r.henadez@company. com)

Happy Christmas to you too.

Best regards

Raul

6.2 Consider inserting your answers within the body of the sender's email (cont.)

Alternatively you could insert your replies into Peter's text:

VERSION 2

> The first is a Request for Proposals that I would like you to read and hear your comments on. There is actually 500 word limit and it is currently around 650, so if you could find any way to remove a few words that would be great.

OK

> Also attached is a proposal for the Request for Funding – for some reason I can't find the email addresses of the relevant people in your office in Madrid, so could you possibly forward it to them?

Done

> you mentioned last time we talked that you had a useful contact at a recruitment agency that you thought I should look up, do you think you could send me the name.

Romeo Henandez (r.henadez@company.com)

> If we don't speak before, I hope you have a Happy Christmas!

Happy Christmas to you too!

Note that the word *Done* means that Raul has already forwarded the proposal – it means *I have done what you asked me to do.* If he hasn't done so yet, he could write *Will do.*

The advantages of Version 2 are that you:

1. can considerably reduce the amount you write and thus the number of potential mistakes. Raul has written only seven words compared to the 60 words of the first version.

2. save yourself time in writing and the recipient time in reading.

3. are more likely to remember to answer all the requests. Also your recipient can see your replies in direct relation to his / her requests.

The only possible disadvantage is that because you write much less, it may seem to the recipient that you are in a hurry and want to deal with his / her email as fast as possible – Version 1 is more friendly. However, given the number of emails that people receive and send every day, this is probably a minor consideration.

6.3 Insert friendly comments within the body of the sender's text

You can use the same technique as illustrated in Version 2 (6.2) to insert friendly remarks within the body of an email you have received. Let's imagine that you work for a company in Pisa, Italy. You have just been to Prague to give some seminars. The email below is from the Czech person who organized the seminars for you. You have inserted your comments within her email.

Hi Paolo

I hope you had a good trip back to Pisa.

Unfortunately there was a three hour delay due to fog, but anyway I got home safely.

I just wanted to say that it was good to meet you last week. I thought the demos went very well.

Thank you. Yes, I was very pleased by the way they went too.

Say hello to Luigi.

I will do. And please send my regards to Lenka Blazkova.

Thank you once again for organizing everything and I hope to see you again in the not too distant future.

Best regards

Hanka

6.4   Saying 'thank you'

When someone has done something for you via email, should you say 'thank you'?

Receiving an email that only says 'thank you' can be quite annoying as it forces you to open and read an email with no real information.

However, if you have received a file from someone and you write back 'thank you' this is a way to acknowledge receipt of the sender's file, so it is actually useful. It also means that the sender does not have to send you an email saying 'Did you get the file?'.

In any case, all you need to do is to add 'thank you' to the existing subject line.

7 CHASING AND BEING CHASED

7.1 Be diplomatic when sending reminders

Ensure that you send reminders in a friendly tone with no sense of frustration or anger. Here are some examples:

> I was wondering if you had had time to look at my email dated 10 February (see below).

> Sorry to bother you again, but I urgently need you to answer these questions.

Empathize with the recipient by showing that you appreciate they are probably very busy.

> I know that you are extremely busy, but could you possibly ...

> I know you must be very busy but if you could find the time to do this ...

> I know this is a lot to ask, but I really need an answer by today.

If you are just reminding a work colleague in the usual everyday routine, then you could say:

> Sorry to bother you but ...

> Sorry to chase you but is there any update on ...?

> Did you get a chance to look at this? [when you have previously emailed an attachment]

7.2 In your reminder, include your original email

When you chase someone with regard to a previous email that you have sent, always include the old email within your new email – this is why in the first example sentence in 7.1 the sender has put *see below*. This indicates that the old mail is below his / her signature of the new email.

A. Wallwork, *Email and Commercial Correspondence,*
Guides to Professional English, DOI 10.1007/978-1-4939-0635-2_7,
© Springer Science+Business Media New York 2014

7.3 Explain the reason for your urgent need for a reply

You may get a reply more quickly if your recipient understands the urgency of your situation. Examples:

Sorry to ask again but any initial feedback on …? – we are very anxious to hear any news.

Any update? It's vital we get moving on the … otherwise we risk … and that would mean that we could lose the …

The thing is – I need to organize my trip to London by early this afternoon.

I was wondering if you had had time to think about my suggestion for a meeting (see email below). If you do think it is a good idea then we would need to arrange it by the end of this week.

I am very much aware that you have a huge workload and I certainly don't want to waste any of your valuable time but I am sure you will appreciate that I need to get the sales figures in by the end of this month. Thanks for your patience.

7.4 Tell your recipient what the new deadline is

Particularly if the situation is urgent, it is a good idea to tell your recipient what the new deadline is. Below are some examples:

I really need to know by tonight whether you will be available in NY in the week commencing Nov 16.

I know this is asking a lot, but I really need an answer by 15.00 London time tomorrow.

I am rather concerned about the situation with the sales presentation, as I urgently need to start going through what you have done on it. If you still need time to work on it, then we need to find a solution together. In any case, I would be grateful if could you get back to me in the next 2–3 hours.

7.5 Motivate your recipient to reply

It may help to motivate your recipient to reply if you do one or more of the following:

- empathize with the fact that they are a busy person who probably has more important things to do than to reply to your request

- explain why this person is important for you and your work

- give them a brief explanation as to why you need a reply so urgently

- tell them how long it will take them to fulfill your request – people tend to overestimate the time that it will take them to complete a task that they don't want to do

- if time is running short, reduce your original request to what is absolutely essential for you (e.g. maybe originally you asked someone to read a whole document, now you just ask them to read just one section)

- find a benefit for them of fulfilling your request

- give them a deadline for their response

7.6 End the reminder with a further apology

You may find it effective to end your email with a phrase such as:

Once again, sorry to have had to bother you with this. Thanks in advance.

Sorry to hassle you, but this is quite an important issue. Any help you could give me would be greatly appreciated.

I apologize for having to ask you this, but I am sure you can appreciate my situation. Thank you.

7.7 Chasing a supplier

Whoever you are dealing with and whatever their level in a company's hierarchy they deserve your respect. When chasing a supplier, quote the order number and date. If this is the first time you are chasing this particular order you can say:

> I wonder if you could help me with a problem.
>
> On April 3 we ordered 800 xTp cartridges (order No. 234 / 3 Apr).
>
> I am just writing to check whether there is any news on the delivery, which I was expecting yesterday.
>
> Anything you could do to speed the process up would be very much appreciated.
>
> Thank you very much in advance.

The recipient of the above email is more likely to be motivated to act on it than if you had used an aggressive and sarcastic tone. Also, remember that writing in another language often acts as a filter and you may not be able to judge the tone of what you have written or the reaction you might receive.

If your first attempt has no success, then you can try a stronger tone in the second attempt.

> On April 3 we ordered 800 xTp cartridges (order No. 234 / 3 Apr) and on April 12 I emailed asking you for an update (see email below). It is now April 14 and there is still no sign of the order.
>
> *As I am sure you can imagine,* we now urgently need the cartridges *and I would kindly ask you to* contact me by the end of today with a firm delivery date. Otherwise, I am afraid we will have to cancel the order.
>
> *Thank you in advance.*

The above email makes the writer's point clear but without resorting to anger and insults. To make it stronger you could remove the phrases in italics.

7.8 Replying to a reminder

The following are examples of replies to reminders. The first three are informal replies to colleagues, the other example is much more formal.

I'm on it. I may be able to send you something *by later today.*

We'll get back to you *when the process is complete.* Hope that is OK.

Sorry, I was quite busy today so I did not get the chance to look in to it. I *may not be able to get back to you before mid next week* – apologies.

Please accept my apologies. I have been inundated with work and had hoped to get back to you last week. Rest assured that I will do my utmost to reply with the information you requested *by Friday at the latest.*

In any case, it is helpful to your recipient if you tell them exactly when you will be able to fulfill their request (highlighted in italics in the examples above).

8.1 Give explicit instructions about how you want the recipient to review your work

When you ask someone to review a report, manual or other written document:

- ensure you ask them politely

- be 100 % explicit exactly what you expect the person to do and tell them what to focus on

---

Dear Carlos

Attached is the report to be presented at the meeting on October 25. It is 25 pages long, but in double line spacing.

I am sending you a Word version, so that you can make the changes directly using Track Changes – please don't use the Comments function.

I know that this is a particular busy time of year for you, so if you can't find the time to read it all, then please focus on Sections 2 and 3, as these are the two sections that have been revised the most since the original draft.

If you could get your revisions back to me by Monday October 21 that would be great.

Could give me a quick confirmation that you will be able to carry out the review.

Thank you very much in advance.

Maria

Note how Maria:

- gives a brief overview of what kind of document it is, and the number of pages (so that the recipient has an idea of the workload involved)

- makes it easy for the recipient to make comments (by sending a Word version of the report rather than a pdf)

- gives the recipient precise details of the parts of the report that most need the recipient's attention

A. Wallwork, *Email and Commercial Correspondence*,
Guides to Professional English, DOI 10.1007/978-1-4939-0635-2_8,
© Springer Science+Business Media New York 2014

8.1 Give explicit instructions about how you want the recipient to review your work (cont.)

- informs the recipient when she needs the report returned (she avoids using a formal and totally unhelpful expression such as *Please could you return it at your earliest convenience*)
- acknowledges that the recipient may be busy
- asks for confirmation from the recipient that he will be able to do the task (if the recipient does not accept, then Maria at least has time to find an alternative solution)

The recipient thus has all the information he needs and thus does not need to ask for clarifications.

8.2 The advantages of using a 'soft' approach

Imagine you wrote the following request:

Hi, Attached is the procedure that we currently use at the end of each year when reviewing the performance of employees. My colleagues in HR implemented this procedure three years ago and feel that it is perhaps time to update it. Could you possibly spare a few minutes to go through it and send us your comments? Your cooperation is very much appreciated.

Decide which of the two replies below you would prefer to receive.

REPLY 1

After having gone through the annual review process, I have some observations and recommendations for you to consider which will make it easier as well as more effective and informative for both the reviewer and the reviewee in the future.

1. The whole process is cumbersome. It could and must be made more straightforward.

2. Abandon the numeric attributes rating system. But if we must continue using this system then consider these amendments: a).... b).... c).... d).... e)...

3. Forms should be made simpler and more user-friendly. I found them almost impossible to work with and had considerable trouble printing them.

REPLY 2

I appreciate the effort you have made in the annual review process. I just have a couple of observations and recommendations that you might find will make the whole process a little more straightforward.

- I am not convinced about the numeric attributes rating system. I found this site (www.howtoreviewemployees/alternatives) which has some great ideas for alternatives to the 1–5 rating system.

- I think it would make your life easier if the forms could be made simpler and more user-friendly (also in terms of printing). One way might be to invest in a professional who can design a simpler and more comprehensive form and / or model it after something that already exists.

Hope this helps.

8.2 The advantages of using a 'soft' approach (cont.)

In each case the recommendations are identical, but the tone of Reply 2 is much more friendly and helpful. Note how the writer of Reply 2:

- begins by saying something positive

- minimizes the seriousness of the criticisms by saying *I just have a couple of observations*

- does not number her points but uses bullets and makes the situation seem less dramatic by only having two points (Reply 1's first point has been incorporated into the introductory part of the email)

- does not use any imperatives or strong modal verbs such as must, have to or should, all of which sound like obligations. Instead she uses other modals might, could and would, which sound like helpful suggestions

- makes helpful suggestions (i.e. the website in the first point, and investing in a professional in the second point)

The result is that the recipient of Reply 2 is not going to feel that in the last three years the review process has been a disaster. He or she is also more likely to be willing to implement the changes suggested in Reply 2 because they have been presented in such a positive way.

I believe that even if you are high up in the company hierarchy using the approach of Reply 2 will always be more effective if your aim is to motivate rather than demotivate your employees.

8.3 Carefully construct and organize your comments on your colleague's work

Below is a possible structure for writing comments on someone's work.

1. Begin with a friendly comment. Try to create a bridge between you and the recipient in which you show that you want to be helpful and cooperative.

   Thanks for sending me your presentation. It's looking really good, well done!

   I enjoyed reading your report. It contains a lot of really useful data. I am impressed!

2. Identify those parts of the document that you feel work really well and give a specific comment saying why you think they work well. By saying something positive, you are putting the reader in the right frame of mind for accepting any criticisms that you may wish to make.

   The introduction is really clear and a helpful overview of the whole project.

   The graphs are cool – what software did you use to generate them?

3. Tell the author of the document what you have done.

   I have read the document carefully and made several changes to the text, including a couple of additions. I hope that in doing so I have not altered the sense of what you wanted to say. In any case, please fee free to disregard.... Where possible, I have tried to... Nevertheless, I think, the doc still needs some work before you send it to the client.

4. When making criticisms be constructive. Rather than saying anything very negative think about whether you could not transform your criticism into a query. For example, instead of saying *I don't understand why you have included the table on x – it seems completely redundant*, you could be more diplomatic and say *Could you clarify why you have included the table on x*?

5. Make any suggestions in a soft way.

   Have you thought about...?

   It might be a good idea to...

   When I am writing a report like this, I find it useful to...

   The client might appreciate it if you...

   I think the boss might be concerned about...

6. Offer further help and tell them when you would be available.

   If you need any more help, then don't hesitate to contact me. I am on vacation next week, but will be back the week after.

   I would be happy to talk through the changes I suggested to the last five slides.

8.4 Learn how to make positive comments

Positive comments certainly ensure a much better chance of getting feedback and making good progress. However, be careful how you phrase any initial positive comments. For example, the following phrase could be interpreted as being a little negative.

I have looked through your presentation and think it's quite good. Just a few comments:

The term *quite good* is dangerous in English because it can mean anything from 'really good' to 'sufficient but nothing great'. This is partly due to the fact that words like *quite* very much depend on the intonation they are given when spoken. But of course in an email this intonation is completely lost.

Imagine how you would feel if you received the comments in the first column below. You would probably feel quite discouraged. The comments in the second column immediately put the recipient in the right frame of mind to receive any criticisms that you might have.

NOT VERY ENCOURAGING	ENCOURAGING
Your presentation is OK.	It's looking really good – I love the way you've used photos.
It looks fine.	Overall it looks excellent and the conclusions are very clear.
I looked at your presentation. Here is a list (non exhaustive) of things you need to change:	I've now had a chance to go through the presentation and I thought you might like a few suggestions.
You need to improve the following points in your presentation:	It's pretty impressive, well done. Here are just a few comments which you are welcome to ignore.

8.5 Be constructive in your criticism

If you need to be critical of someone's work, your recipient is more likely to act on your comments if they are presented in a constructive way.

Let's imagine that you are co-authoring a report on how to reduce the company's travel costs. However, the colleague you are co-authoring with is from another branch of your company in another country. You and your colleague are on the same level in the company hierarchy. Most of your communication has been conducted via email and you have only met once face to face.

In your opinion your co-author has committed three mistakes. He has:

1. forgotten one area of the travel costs: car rentals

2. not put the sources of his statistics on travel costs between Asia and Europe

3. misspelled the names of some of the airlines

Here is a typical example of an email that fails to address these points in a constructive way.

> Dear Paul
>
> I have had a look at the report and there are several problems with it. First you have failed to take in account car rental costs. Second, you have omitted the sources of your stats on costs between Asia and Europe. Last but not least, you have misspelled a number of the airline companies.
>
> I am reattaching the draft with various other suggested changes and additions.
>
> Please could you make the other necessary changes and send me the draft back by the end of this week. It is now quite urgent.
>
> Best regards
>
> Maria

Maria has not thought about how Paul might feel when he reads her email. Paul is likely to be very angry and / or very upset. He may have spent several days writing the report. In addition, there may be reasons for the three apparent errors. Perhaps the car rentals costs were originally in the report, but Paul had decided to change their position, and had thus cut them out but forgotten to paste them back in (maybe he was interrupted by a phone call when in the middle of the cut and paste). In his email to Maria,

8.5   Be constructive in your criticism (cont.)

perhaps Paul had forgotten to tell her that he was locating the sources of some of his statistics. And finally, maybe Paul was waiting for Maria's revisions before doing a final spell check.

Maria's email will have a negative impact because:

- it does not give the recipient the benefit of the doubt, it assumes that there is no other explanation for Paul's three errors other than that he is incompetent (see 9.7)

- it is very direct, there is no introduction

- it over-emphasizes the problems by using words like *failed* and *last but not least*, it thus seems quite sarcastic

- its tone is not that of a colleague but that of an angry boss

If you begin your email with an immediate criticism, your recipient will react negatively and this will set the tone for how he / she reacts to your other comments.

Instead find something positive to say. Here are some examples:

Thanks for getting this report to me far ahead of the deadline, this will make my life much easier.

I really like your succinct style of writing, I think it will help the readers to digest the report quickly and easily.

Provided they are sincere, the above sentences will help to get the recipient in a positive state so that he / she will then read your criticisms with a more open mind. You then have to deal with the three mistakes in his / her draft. The first thing is to try and reduce the number of criticisms you make, and then to begin with the most serious mistake. Thus Maria's email could be rewritten as follows:

I just wanted to point a couple of issues in your draft.

I may have simply not seen them, but I couldn't find any data on car rentals.

Secondly, the boss will expect us to provide the sources for all our statistics, so I think we need to add these. I think I only noticed a couple of cases, so this shouldn't take you too long.

By the way, would you mind doing a final spell check, but not just with Word as I don't think it will identify any spelling mistakes in the technical names (e.g. the names of the airline companies).

8.5   Be constructive in your criticism (cont.)

In her revised version, it seems that she is only making two criticisms
(*firstly… secondly*) and she does this in a very soft indirect way. This gives
Paul the option to prove her wrong. Regarding the sources of the statistics,
Maria takes joint responsibility for this by using the inclusive pronoun *us*
and *our*, and she then minimizes the effort required by referring to the
small number of sources involved and the short time required to sort out
the problem. Finally, she presents the reference to the spelling mistakes
almost as an afterthought (*by the way*) and as a friendly request (*would
you mind*).

So, when Paul reads the above message, he will be appreciative of
Maria's comments and will be more likely both to implement them and get
back to Maria quickly.

8.6 Avoid being too direct when asking for clarification and making suggestions

When you are not sure about something that your colleague has written, it's a good idea to be careful how you ask questions. Also, avoid being too direct otherwise you could sound more critical than you intend. You can make a question softer by:

- adding a short introductory phrase

- making out that it's you that has a difficulty, but not necessarily that this difficulty is caused by some negligence on the recipient's part

- phrasing the question in the passive, thus making it appear that the recipient was somehow not involved in the decision when in fact he / she was

DIRECT QUESTION	INDIRECT QUESTION
Why don't you have an 'Outline' slide?	By the way, have you thought about having an 'Outline' slide?
When are you going to mention the disadvantages of our approach?	Have you decided when you are going to mention the disadvantages of our approach?
Why did you include the table in the fourth slide?	It was probably my idea, but can you remind me why it was decided to include the table in the fourth slide?

Of course, if you have a lot of comments to make or if your comments are simple observations that will clearly help to improve the content of the presentation, then there is no need to always adopt a roundabout approach. For example:

The weight should be quoted to 3 decimal places, not 4.

An easier solution would be to swap the position of slides 5 and 7.

Don't forget to do a spell check at the end (I always forget!)

In the above cases you are not criticizing the recipient but merely making some helpful suggestions.

8.7   Conclude by again saying something positive

Your concluding comments should always be positive, thus leaving the recipient thinking that they have not made a complete disaster of their work. Make sure you don't simply end with *Regards,* but include a phrase like one of the following:

Thanks for your hard work on this. Much appreciated

Well, I think that's all – once again, a really excellent job, just a few things to tighten up here and there.

Hope you find these comments useful, and bear in mind that I've only focused on what changes I believe need making so I'm sorry if it comes across as being very critical.

8.8   Re-read everything before you hit the 'send' button

Always re-read what you have written when you have been criticizing someone's work (or whenever you have something potentially difficult to say). Make sure you haven't said anything that could damage your relationship or offend in any way (see also 9.7). Also, if you really have to be critical, consider leaving the email for a while and then coming back to it to see if you have been fair or not.

8.9   Responding to criticism

It pays to be appreciative of other people's input on your work. This is irrespective of whether their comments were useful to you or not, and of whether you agree with what they say.

You can begin an email of thanks by saying:

Thank you so much for your review. The report has certainly benefited from your input – particularly the conclusions, where you have managed to really highlight the…

If they have made any specific criticisms you can say:

I understand what you meant by… so I have adjusted that section accordingly.

Having read your comments, I now realise that I need to rewrite the part about…

I think you were right about the table, so I have…

If you need any clarifications you can say:

Thanks very much for all this. Just one thing – could you just clarify exactly what you mean by …

I may come back to you if I need further thoughts on some of the slides.

9.1 Think about how your email might be interpreted by the recipient

When you write in English, you may be less worried about how your email might be interpreted than you would if you were writing to a colleague of your own nationality. For many non-native speakers writing in English is like writing through a filter: the way you write seems to have much less importance than it would if you were writing in your own language.

Below is an email from one colleague to another. They are on the same level in the company hierarchy.

> Here is a draft of the manual. Read and check everything: in particular, you have to work on the introduction and prepare the screenshots.
>
> You should send it back to me by the end of this month at the latest.
>
> I remind you that this manual is for a very important client.

For a native English speaker the above email sounds like a series of orders given by someone very high in a hierarchy. Thus the recipient might be a little surprised or offended by the tone. The problems are due to the use of:

- the imperative (*read and check*) – this gives the impression that the sender is not on an equal level to the recipient, but rather quite an aggressive boss

- *have to* – this sounds like a strong obligation rather than a request

- *should* – again, this sounds like an order

Although in your language the use of the imperative and *have to / should* may be perfectly acceptable in this context, in English they are not.

A. Wallwork, *Email and Commercial Correspondence,*
Guides to Professional English, DOI 10.1007/978-1-4939-0635-2_9,
© Springer Science+Business Media New York 2014

9.1 Think about how your email might be interpreted by the recipient (cont.)

The email could be improved as follows:

> Here is a draft of the contract. Please could you read and check everything. In particular, it would be great if you could complete / revise the introduction and also prepare the screenshots.
>
> Given that our deadline is the second week of next month, I would be grateful to receive your revisions by the end of this month.
>
> As you know this manual is for a very important client – so the boss wants us to do a really good job!

But using the imperative form is not always impolite or inappropriate. For example, when you are giving a formal list of instructions these will generally be quicker and easier to follow in the form of imperatives (i.e. the infinitive form of the verb without *to*). So the first rather than the second sentence below would be more appropriate in a list of instructions:

Attach your application form to your email.

The application form should be attached to the email.

This approach will not be rude:

* if you have a friendly introductory phrase before a list of commands

* if the rest of the mail is friendly

If in doubt, use *please*.

9.2 Use non-aggressive language

Whenever you write an email, always be aware that there is probably more than one way to interpret what you have written. Before you send your email, check for potential misinterpretations, and rewrite the offending phrase.

For example, here is what appears to be an inoffensive reminder.

> For your information I remind you that it is VERY important to always specify your current workstation IP address.

However, this sentence has various problems:

- *For your information* could be interpreted as sounding like someone who has been contradicted and is now giving their point of view in quite an aggressive way

- *I remind you* – the present tense in English is sometimes used to give a sense of authority or formality. It thus sounds very cold and unfriendly

- *VERY* – rather than using capitals, consider using bold

Here are two different ways of rewriting the sentence:

> Just a quick reminder – don't forget to specify your current workstation IP address. Thanks!

> I'd just like to remind you that the IP address of a workstation must always be specified.

The first alternative is informal and friendly. The second is more formal, but uses three tricks to make it soft:

1. a contracted form (*I'd* rather *than I would*) which gives the phrase a less authoritarian tone

2. the passive form – this then makes the *IP address* the subject of *specified* (rather than the implicit *you must specify*)

3. *a workstation* rather than *your workstation* – this makes the message sound that it is not directed personally at the recipient

Below are some more examples:

AGGRESSIVE	NON AGGRESSIVE
You have sent me the wrong attachment.	I think you must have sent me the wrong attachment.
I need it now.	I appreciate that this is a busy time of year for you but I really do need it now.
I have not received a reply to my email dated…	I was wondering whether you had had a chance to look at the email I sent you dated… (see below)

9.2 Use non-aggressive language (cont.)

In summary: Use a more indirect, softer approach and include an introductory phrase that in some way tones down any aggression. If possible empathize with your reader's situation.

9.3 Avoid irritating the recipient with unnecessary remarks

When you revise your email before hitting the 'send' button, make sure you remove anything that is not strictly necessary, particularly phrases that might annoy the recipient. Recipients do not like to be treated like schoolchildren or be made to feel guilty, thus in most contexts the phrases below should be deleted:

This is the second time I have written to request...

I am still awaiting a response to my previous email...

As clearly stated in my previous email,

If you had taken the time to read my email carefully...

9.4 Choose the most appropriate level of directness

Below is an example of a request written in various ways from very direct (using an imperative) to extremely polite. You can choose the one you feel is the most appropriate:

Revise the document for me.

Will you revise the document for me?

Can you revise the document for me?

Could you revise the document for me?

Would you mind revising the document for me?

Do you think you could revise the document for me?

Would you mind very much revising the document for me?

If it's not a problem for you could you revise the document for me?

If you happen to have the time could you revise the document for me?

When you translate from your own language into English you may lose the sense of politeness that the version in your own language had. Thus it is possible that an email that in your language would sound courteous may, when translated into English, sound quite rude.

9.5 Be friendly but not invasive

When you call someone on the phone, you probably begin by asking *how are you?* You are not necessarily interested in the answer, but it is just a formality at the beginning of a phone call. Some people also ask this question at the start of an email – again they may not be expecting an answer, but it just acts as a friendly start rather than being too direct.

If you have a good relationship with your recipient then they are more likely to carry out your requests and do so more quickly than they might if you are totally anonymous to them. One tactic after a few email exchanges is to reveal / announce some personal information.

This could be at the beginning of the email, for instance:

Hope you had a good weekend. I spent most of mine cooking.

So how was your weekend? We went swimming – we were the only ones in the sea!

How's it going? I am completely overloaded with work at the moment.

Or it could be the end of the email:

Regards from a very hot and sunny Caracas.

Hope you have a great weekend – I am going to the beach.

These little exchanges only take a few seconds to write (and to be read, i.e. by the recipient). Also, by making comments such as these, you might discover that you have something in common (cooking, swimming) and this will give you something to 'talk' about in your emails.

Such exchanges 'oil' the relationship – but of course you must also be sincere. The result is that any future misunderstandings are likely to be resolved more quickly and with a better outcome than there might be with an anonymous interlocutor.

However, it is really important not to take this to extremes. For instance, never write:

I saw your status on Facebook. It seems you had a nice time in Venice!

Although Facebook is public and was specifically designed to let people into your private life (or at least that part of your private life that you want them to have access to), some users of Facebook find the idea of people who they hardly know looking at their pages and then commenting on them as being quite distasteful. It is a bit like being stalked. So, be careful to respect people's privacy and not be invasive.

9.6   Add a friendly phrase at the end of an email

There are various phrases that you can use at the end of an email, particularly if you think the rest of the email may be a little strong. These include:

Have a nice day.

Have a great weekend.

Keep up the good work.

9.7   Avoid labeling your recipient as 'incompetent'

When we don't get the right or appropriate answers to questions that we have asked in an email, there is a natural human tendency to consider the recipient as being incompetent at their job. This inevitably leads to friction in our email relationship with this person.

The secret is to think of the recipient as the person sitting next to you, i.e. a normal helpful hard-working colleague who has to deal with many other emails – not only yours.

If you are consistently getting the wrong answer to your questions, it may be that you are:

• not phrasing your questions or instructions clearly

• addressing your questions to the wrong person

In the first case you could write to your recipient:

I am concerned that I am not getting the appropriate answers to my questions, and I suspect that it may be because I am not expressing myself clearly. Below, I have rephrased my questions. Please let me know whether the questions are clearer and whether / when you will be able to provide me with answers. Thanks very much.

In the second case you could write:

I am concerned that I am not getting the appropriate answers to my questions, and I suspect that it may be because you are not actually the right person to contact. If this is so, then please could you give the address of the right person to contact. Sorry to bother you with this, but I need to get answers as soon as possible. Thanks very much.

Obviously nothing guarantees that after adopting one of these strategies you will get the answers you want. If this is the case, then your best option is probably to telephone the person directly.

10 RECOGNIZING LEVEL OF FORMALITY

Most people try to match the level of formality of the email that they have received. But this entails knowing how to recognize just how formal an email is. There are various clues.

10.1 Formal: long and complex sentences

If a phrase is long and / or complex, this is generally a sign of greater formality.

FORMAL	LESS FORMAL
We *have pleasure in confirming* the acceptance of your order for …	*This is to confirm* that your order has been accepted for …
Should you need any clarifications please do not hesitate to contact us.	*If you* have any questions please let us know.
You are requested to acknowledge this email.	*Please* acknowledge this email.
It is necessary that I have the report by Tuesday.	*Please* could I have the report by Tuesday.

It is important to be aware, however, that some short sentences are not always the most informal and can also come across as rather cold. Writing in a telegraphic style can obscure the meaning from your reader, so always try to write complete and comprehensible sentences.

The examples below show how a simple concept, such as acknowledging receipt of a mail, can be expressed in many different ways: from completely detached (the first example) to quite warm.

I confirm receipt of your fax.

This is just to confirm that I received your fax.

Just to let you know that your fax got through.

Thanks for your fax.

A. Wallwork, *Email and Commercial Correspondence,*
Guides to Professional English, DOI 10.1007/978-1-4939-0635-2_10,
© Springer Science+Business Media New York 2014

10.2   Formal: modal verbs

The modal auxiliaries *may, can, could* and *would* are often used to make a request sound more courteous and less direct. Compare:

> *May* I remind you that we are still awaiting receipt of our order No. 1342/2.

> We are still awaiting receipt of our order No. 1342/2.

> *Can* you kindly check with her that this is OK.

> Check that this is OK.

> *Could* you please keep me informed of any changes you plan to make to the presentation.

> Keep us informed of any changes you plan to make to the presentation.

> *Would you like* me to Skype you?

> Do you want me to Skype you?

In a similar way, *won't be able to* is often preferred to *cannot*, and *would like* or *wish* to *want*. Both *cannot* and *want* tend to sound too abrupt.

> I'm sorry but I *won't be able to* give you any feedback on your report until next week.

> We regret to inform you that we *will not be able to* offer any special rate to delegates from your company.

The modal verb *may* is extremely useful whenever you want to give your mail a formal tone:

> I would be grateful for any further information you *may* be able to give me about ...

> *May* I thank you for your help in this matter.

Note: The use of *shall* as a future auxiliary and *should* as a conditional auxiliary is outdated in English, and their use is a sure sign of formality. In the examples below, the first sentence in each pair is very formal, the second sentence is normal English.

> We *shall* give your request our prompt attention.

> = We *will* deal with your request as soon as possible.

> I *should be glad if you could* send the file again, this time as a pdf.

> = *Please could you* send the file again, this time as a pdf.

10.3 Formal: nouns

When there is a predominance of nouns rather than verbs this gives an email a feeling of distance and formality:

Please inform me of the time of your *arrival*.

Please let me know when you *will be arriving*.

To the best of our *knowledge*.

As far as we *know*.

10.4 Formal: multi-syllable words

Generally a clear indication of formality is given by the number of syllables in a word – the more there are, the more formal the email is likely to be. If you speak French, Italian, Portuguese, Romanian or Spanish, a good tip is if the multi-syllable word in English looks similar to a word that you have in your own language, then it is probably formal in English. Compare the following pairs of verbs. The first verb is multi-syllable and formal, the second is monosyllable or a phrasal verb.

FORMAL	INFORMAL	FORMAL	INFORMAL
advise	let someone know	evaluate	look into
apologize	be sorry	examine	look at
assist	help	inform	tell
attempt	try	perform	carry out
clarify	make clear	receive	get
commence	start	reply	get back to
consider	think about	require	need
contact	get in touch	utilize	use

The same also applies to nouns, e.g. *possibility* vs *chance*.

10.5 Omission of subject and other parts of speech

A clear sign that an email is informal is when the subject of the verb and / or the auxiliary are missing. An email is even more informal when articles, possessive adjectives etc are also missed out in telegraphic style (last example).

INFORMAL	FORMAL
Been very busy recently.	*I have been* very busy recently.
Appreciate your early reply.	*I would appreciate* your early reply.
Hope to hear from you soon.	*I hope* to hear from you soon.
Speak to you soon.	*I will speak* to you soon.
Looking forward to your reply.	*I am looking* forward to your reply.
Will be in touch.	*I will* be in touch.
Just a quick update on …	*This is just* a quick update on …
Have forwarded Carlos *copy* of *ppt* to *personal* email too.	I have forwarded Carlos *a copy* of *the presentation* to *his personal* email too.

10.6 Abbreviations and acronyms

Some abbreviations are perfectly acceptable even in a formal email, such as *re* (regarding) and *C/A* (bank current account). Others however such as, *ack* (acknowledge, acknowledgement), *tx* (thanks), *rgds* (regards) should be used with caution – they could give the impression that you could not find the time to write the words out in full. For more on this topic see 15.2 and 15.7.

10.7 Smileys

A smiley is a clear indication of informality. I strongly suggest that you use them only if your recipient has used them first, as there are some people who find smileys annoying. Also, avoid using them with anyone when you want to make a difficult request seem lighter. For example:

Please could you send me the revised draft tomorrow :)

The above request for someone to review a long document within a very short timeframe is not helped by having a smiley, which may actually annoy the recipient as he / she will certainly not be happy to do such a long task in such a short time.

To see a list of smileys: 15.9

10.8 Avoid excessively formal forms of English

Over the centuries and particularly in the last few decades, the English language has increasingly become more and more informal. Below are three examples of salutations from letters written by Benjamin Franklin, one of the founding fathers of the USA, in the late 18th century.

Your faithful and affectionate Servant,

I am, my dear friend, Your's affectionately,

My best wishes attend you, being, with sincere esteem, Sir, Your most obedient and very humble servant,

Such phrases today would sound ridiculous in an email, even in a very formal letter. However, similar phrases exist in many languages of today. For example, phrases such as *Would you accept, sir, the expression of my distinguished salutation* (10 words), or *In expectation of your courteous reply, it is my pleasure to send you my most cordial greetings* (17 words), sound extremely pompous in English. They would probably be rendered as: *I look forward to hearing from you* (7 words) or simply *Best regards* (2 words).

In fact, most languages in their written form tend to be more formal than written English. This formality shows itself not just in the choice of words and expressions but also in the length of sentences and paragraphs. Below is an email to a human resources manager from someone who wishes to apply for a job. The parts in italics would be considered much too formal (or simply wrong) by most Anglos. The best solution is to omit them.

Dear Human Resources Manager,

With due respect I would like to draw your attention that I have a degree in ... I am *highly* interested in continuing my research in the field of reducing fuel emissions and thus of working in your *esteemed* research and development team. I am sending *herewith* my bio-data *in favor of your kind consideration.*

I would very much appreciate it if you would consider me for a position in your R&D division.

I am eagerly looking forward to your generous suggestion.

With warmest regards.

Sincerely yours

In any case, NEVER translate a standard phrase from your own language into English. Instead, use an equivalent phrase (see the useful phrases in Chapters 16 and 17).

10.9   Don't mix levels of formality

Below is an email written by one business person to another. They had previously met at a trade fair. There is a strange mix of informal (in italics) and formal. This mix may make the recipient feel that the sender is unprofessional.

Dear Miroslav,

I hope you *have been having a really good time* since our meeting at the trade fair in Belgrade.

I would be very grateful if you could kindly tell me how to obtain the DS2019 document in order to request the visa.

I would like to thank you in advance and *have a great Xmas.*

Cheers,

Lamia Abouchabkis

10.10 Avoid very colloquial English

In your email correspondence you may learn many new phrases from native English speakers. However, there is a chance that if you use them yourself:

- you may use them incorrectly

- your non-native recipient may not know what they mean

In the table below are some typical phrases (in italics) used by native speakers with their more formal English equivalents.

COLLOQUIAL ENGLISH	EQUIVALENT IN NEUTRAL ENGLISH
If you need any more info then just *shout*.	let me know
If you can *start the ball rolling*.	get things started / initiate the process
Just so that we *are all on the same page* ...	all have the same information
Would you mind *bringing me up to speed*?	updating me / telling me the most recent developments
Hi, I just wanted to *touch base*.	get in touch with you to let you know the current situation
Do you think is this *doable*?	can be done
I'll *keep you posted*.	keep you up to date
I'm just interested in a *ballpark figure*.	a rough numerical estimation

11 PUNCTUATION AND CAPITALIZATION

11.1 No punctuation necessary after salutations

You don't need to put any punctuation after your initial salutation or in your final salutation.

Dear Mr Lee

This is to inform you that your invoice has now been paid.

Best regards

Reza Bahram

Note that in the above email:

- the words *you* and *your* have no initial capitalization. In English, the forms *You* and *Your* to show respect do not exist
- the first word of the first line (*This*) of the body of the email has an initial capital letter

This means that the following email is NOT correct:

Dear Mr Lee

this is to inform You that Your invoice has now been paid.

best regards

Reza Bahram

A. Wallwork, *Email and Commercial Correspondence,*
Guides to Professional English, DOI 10.1007/978-1-4939-0635-2_11,
© Springer Science+Business Media New York 2014

You can if you wish use a comma after a salutation, so the above email could also be written as follows:

11.1   No punctuation necessary after salutations (cont.)

Dear Mr Lee,

This is to inform you that your invoice has now been paid.

Best regards,

Reza Bahram

If you follow the rules given in this chapter for writing clearly and concisely, you will generally only need to punctuate your email with commas (,) and full stops (.). In an email you never need to use semicolons unless you are dividing the items in a list.

11.2 Hyphens

Hyphens (-) are frequently used instead of commas to add afterthoughts and comments:

The meeting has been rearranged – again!

Thanks for this – it is looking great.

11.3 Exclamation marks and smileys

Exclamation marks (!) and smileys (see 10.7) can both be quite risky as the reader can never be sure of the tone intended:

• does your exclamation mark mean you are annoyed or expressing your sense of humor?

• does a smiley really mean you are happy or are you being sarcastic?

If possible the actual words of the email should be unambiguous in tone. If you think there might be more than one possible interpretation, and that one of these misinterpretations might offend, aggravate or upset the recipient, then either rewrite the sentence or make a phone call instead.

11.4   All caps

Do not use all caps within the main text of an email.

How would you feel if you received this email.

> Hi
>
> HAVE YOU FINISHED THE REPORT YET? I NEED IT URGENTLY. IN ADDITION, I NEED YOU TO GIVE ME DETAILS OF XYZ.
>
> BLAH BLAH BLAH

Firstly, reading any text which is all in capitals is more difficult than reading upper / lower case text.

Secondly, the use of all caps could be interpreted as expressing the writer's frustration or anger, and thus will probably have a negative impact on the reader.

However, using 'all caps' is useful to highlight headings. For example:

> Dear all
>
> Below is the agenda for tomorrow's meeting.
>
> TIMEKEEPING
>
> Several employees seem to be arriving late on a regular basis. This is causing friction amongst ...
>
> DEADLINES
>
> In order to keep customers happy, we need to improve our ability to respect deadlines – particularly given that we are the ones who actually set the deadlines. In addition, ...
>
> HOLIDAYS
>
> Given the impending project deadlines, all holidays are to be suspended until ...

Bold can also be used for the same purpose.

12 SENDING ATTACHMENTS

12.1 Consider not sending an attachment to someone with whom you have had no previous contact

Some people do not appreciate receiving attachments from people with whom they have had no previous contact. You can avoid sending attachments by giving the recipient a link where, if they wish, they can download your document.

12.2 Always tell your recipient when you have attached a document

Be explicit in stating that you have attached a document, otherwise your reader may not notice. You can say:

Please find attached sales forecasts for next year.

Attached are sales forecasts for next year.

Use the word *attached* rather than *here is / are*. If you use *here is / are* there is a risk that the reader will think the details are contained within the email itself. Example:

Here is the sales forecast for next year.

A. Wallwork, *Email and Commercial Correspondence,*
Guides to Professional English, DOI 10.1007/978-1-4939-0635-2_12,
© Springer Science+Business Media New York 2014

12.3 Detail any changes with respect to a previous document

If you are sending an updated document with respect to the document that your recipient already has, explain any key differences. Example:

> Dear all,
>
> Please find attached a new version of sales trends.
>
> Besides info on the geographical distribution by country, for some countries details have been now been added on sales in particular cities.
>
> Also there are now new reports in a preset layout, some of which are still drafts.
>
> It would be really helpful if you could let me know if there are any discrepancies, so that these can then be ironed out in this first version.

If someone has requested you to send them a document, then explain if and how the document differs from the requirements specified by your interlocutor.

> Attached is the current organizational chart. It's a few months out of date, but the management is all the same. Note, however, that since this document was drafted, a new level has been added under ...

12.4 Instruct the recipient on what feedback you expect on the attachment

Below are some examples of emails where the sender is requesting feedback:

> Please find attached a list of items that ... Please let me know if there is anything else you would recommend that I add to this list or anything in particular you would like to know about ...

> Attached is a summary of ... The summary gives an idea of ... If you think it would be useful, it could be circulated fortnightly and if required can be sent to other departments as well. Please let me have your comments, suggestions for future improvements and anything you think might be useful.

> Attached is the list of people that .... Please let me know whether the instructions attached are clear enough. Once you have had a read through, it would be good to get your feedback and discuss how to make the task of updating this spreadsheet as simple as possible.

12.5 For non-work attachments, explain why the attachment will be of interest to the recipient

When you send an attachment that you think might of interest to someone – rather than being an essential work item – explain why you think the recipient might find it interesting. In some cases the name of the attached file might be self explanatory, but if it is not then at least provide a one-line explanation.

In any case, if you want people to open your attachment it pays not to become renown for sending random attachments on a frequent basis. If you do this, people are likely to trash your email immediately without even opening it.

13.1 Note the differences between an email and a business letter

A business letter is only marginally different from a formal email. The main differences are that it:

- looks more official (given the letterhead)

- is designed to be printed

- has a different layout which tends to include more information (e.g. address, date)

- tends to be written in a formal style (see Chapter 10)

- generally contains a real handwritten signature

13.2 Templates

You can find many templates on the Internet. Two of the most frequently accessed are Google Docs Templates and jobsearch.about.com. Some will also help you with the content and some provide templates for letters for almost every possible occasion (e.g. writeexpress.com).

When you look at the templates, see which layout style you think works best.

A. Wallwork, *Email and Commercial Correspondence,*
Guides to Professional English, DOI 10.1007/978-1-4939-0635-2_13,
© Springer Science+Business Media New York 2014

13.3   General rules on layout of letters

Your aim is for your reader to be able to read and absorb the information in your letter as quickly and effectively as possible. A good clear layout out will help you to achieve this aim. There are no set rules for the layout, so here are my guidelines:

- only use one font
- align everything to the left
- no indentations
- have a subject line, center it and put it in bold
- have at least 6 pt space between one paragraph and another
- use bold for emphasis rather than underlining

Alignment to the left makes the letter look clean and easy on the eye. However, if you are hoping to get work from a particular client who aligns addresses, dates and other items to the right, then you may wish to copy their style. Also you may wish to bear in mind clients in countries who typically read from right to left.

13.4 Addresses

Your address: Business letters normally have the address of the company automatically inserted onto the page. If you are writing a personal letter, then the simplest and clearest way is to write the address in one line across the header or footer.

Their address: This should be the first item in your letter and like everything else should be aligned to the left.

If you put the name of the recipient in your initial salutation (e.g. *Dear Anna Southern*) then there is no need to put in the address.

However, if you are writing to someone whose name you do not know then you may wish to precede their address with the job position of this person: You can write:

For the attention of the HR Manager

ABC Inc.

12–18 North Street

Kansas City

MO 64105

USA

The acronym of 'for the attention of' is FAO, so can write:

FAO: Sales Manager

FAO: Invoice Processing

13.5 References

References may refer to orders, invoices, client numbers etc. These are normally located on a separate line either immediately before or after the date.

13.6   Dates

Write dates as follows:

10 March 2030 (no punctuation required)

or

March 10, 2030

but not

~~10 / 03 / 30~~

as the reader may interpret this not as 10 March but as 3 October.

There is no need to use 1st, 2nd, 3rd, 4th or 1st, 2nd, 3rd, 4th etc. These abbreviations are completely unnecessary. Moreover, you may mistakenly use the wrong one!

In the USA and GB, people do not put the location before the date. So the following would normally be considered wrong:

~~London, 10 March 2030~~

~~New York, March 10, 2030~~

13.7   Subject lines

A subject line is useful. It immediately tells the reader what the letter is about.

You simply need to write the subject without any introductory word. Thus, the first example below is good, the others are not.

Your invoice No. 1424, dated 10 March 2030

~~Subject: Your invoice No. 1424, dated 10 March 2030~~

~~Object: Your invoice No. 1424, dated 10 March 2030~~

~~Reference: Your invoice No. 1424, dated 10 March 2030~~

Subject lines are normally centered and appear before or after the initial salutation. Put your subject line in bold. Here is an example.

Dear Ms Lucejko

Your invoice No. 1424, dated 10 March 2030

This is to confirm that yesterday we made a bank transfer for an amount of $45,000 in settlement of your invoice No. 1424.

Best regards

Hao Li

13.8   Initial salutation and final salutation

Salutations follow the same rules as in an email (see Chapter 3). Obviously, in a written letter you are not likely to use informal salutations such as *Hi, Hello* and *Good morning*.

13.9   Body of the letter

Your letter will be read more easily if:

- the paragraphs are short
- each paragraph is separated by white space
- there is no indentation
- the subject line is centered and in bold
- everything else is aligned to the left

Compare these two versions of the same letter. Which layout do you prefer? Which would be quicker to format? Which would be easier to read?

Most people would find the layout in Version 2 easier to format, more aesthetically pleasing and easier to read.

VERSION 1

ABC Consulting
1, The Avenue
London EC 5

10 April 2030

Headcount figures for first quarter 2030.

Dear Sadiq Irfan,

Blah blah blah blah blah blah blah blah blah blah blah blah blah blah blah
blah blah blah blah blah blah blah blah blah blah blah blah blah blah blah blah
blah blah blah blah blah blah blah blah blah blah blah blah blah blah blah blah
blah blah blah blah blah blah blah blah.

Blah blah blah blah blah blah blah blah blah blah blah blah blah blah blah
blah blah blah blah blah blah blah blah blah blah blah blah blah blah blah blah
blah blah blah blah blah blah blah blah blah blah blah blah blah blah blah blah
blah blah blah blah blah blah blah blah.

Blah blah blah blah blah blah blah blah blah blah blah blah blah blah blah
blah blah blah blah blah blah blah blah blah blah blah blah blah blah blah blah
blah blah blah blah blah blah blah blah blah blah blah blah blah blah blah blah
blah blah blah blah blah blah blah blah.

Blah blah blah blah blah blah blah blah blah blah blah blah blah blah blah
blah blah blah blah blah blah blah blah blah blah blah blah blah blah blah blah
blah blah blah blah blah blah blah blah blah blah blah blah blah blah blah blah
blah blah blah blah blah blah blah blah.

I looking forward to hearing from you in the near future.

Best regards,

Simona Rodriguez

VERSION 2

ABC Consulting
1, The Avenue
London EC 5

10 April 2030

Headcount figures for first quarter 2030.

Dear Sadiq Irfan,

Blah blah blah blah blah blah blah blah blah blah blah blah blah blah blah blah blah blah blah blah blah blah blah blah blah blah blah blah blah blah blah blah blah blah blah blah blah blah blah blah blah blah blah blah blah blah blah blah blah blah blah blah blah blah blah blah blah blah blah blah blah.

Blah blah blah blah blah blah blah blah blah blah blah blah blah blah blah blah blah blah blah blah blah blah blah blah blah blah blah blah blah blah blah blah blah blah blah blah blah blah blah blah blah blah blah blah blah blah blah blah blah blah blah blah blah blah blah blah blah blah blah blah blah.

Blah blah blah blah blah blah blah blah blah blah blah blah blah blah blah blah blah blah blah blah blah blah blah blah blah blah blah blah blah blah blah blah blah blah blah blah blah blah blah blah blah blah blah blah blah blah blah blah blah blah blah blah blah blah blah blah blah blah blah blah blah.

Blah blah blah blah blah blah blah blah blah blah blah blah blah blah blah blah blah blah blah blah blah blah blah blah blah blah blah blah blah blah blah blah blah blah blah blah blah blah blah blah blah blah blah blah blah blah blah blah blah blah blah blah blah blah blah blah blah blah blah blah blah.

I looking forward to hearing from you in the near future.

Best regards,

Simona Rodriguez

14 PLANNING AND STRUCTURING AN EMAIL OR LETTER, AVOIDING MISTAKES IN YOUR ENGLISH

14.1 Plan your email or letter and be sensitive to the recipient's point of view

Think about the following.

- What is the goal of my email / letter?

- Who is my recipient?

- What is their position in the company hierarchy? How formal do I need to be?

- How busy will my recipient be? How can I get his / her attention?

- What does my recipient already know about the topic of my email?

- What is the minimum amount of information that my recipient needs in order to give me the response I want?

- Why should my recipient do what I want him / her to do?

- What is my recipient's response likely to be?

Write in a way that shows you understand the recipient's position and feelings. Even though you may be requesting something from them, you are at least doing so by trying to address their needs and interests as well.

Think about how your recipient will interpret your message – can the message be interpreted in more than one way, is there any chance it might irritate or offend the recipient, will they be 100 % clear about what its purpose is?

Also, if possible try to think of a benefit for the recipient of fulfilling your request.

Even if you contact someone frequently, you cannot assume that they will know the reason for your message.

A. Wallwork, *Email and Commercial Correspondence,*
Guides to Professional English, DOI 10.1007/978-1-4939-0635-2_14,
© Springer Science+Business Media New York 2014

14.2 Organize the information in the most logical order and only include what is necessary

The email below is to a manager in a company in South Korea. The sender is informing the recipient of a delay in the shipment of an order.

Dear Mr Kyeong

At the moment we are not able to send the order you requested.

The order number is the following: # 08SFL-00975

We underestimated the time it would take to fulfill your order. This was due to the following reasons:

- aaa

- bbb

- ccc

We would be very grateful to you if we could grant us a delay of a couple of weeks for delivery of the order.

We are confident that we will be able to ship the order by 21 October.

Best regards

The above email is not effective because:

- the most important information (i.e. the new delivery date) is left until the end of the email

- it contains information that may be irrelevant to the reader (i.e. the reasons for the delay)

- there is no apology

It is also a good idea not to force the recipient to read a mass of non-essential information before you finally tell them your request. The above email could thus be rewritten as:

14.2 Organize the information in the most logical order and only include what is necessary (cont.)

Dear Mr Kyeong

We are very sorry but unfortunately we will not be able to ship your order number # 08SFL-00975 until 21 October.

We realise that this may cause you inconvenience and we are thus happy to grant you an additional 5 % discount on the order.

Please let me know if you need any clarifications and once again let me apologize for this delay

Best regards

You may think the above is rather direct, but the recipient will appreciate the fact that he / she only needs to spend three seconds on reading your email.

14.3 Bear in mind that long emails / letters will be scrolled

In long emails and letters it is imperative to gain the recipient's interest quickly so that they will be encouraged to continue reading. Ensure that there is a topic sentence showing what the email / letter is about, and what response or action you require from the recipient.

If you have written a long email / letter, it is generally a good idea to have a bulleted summary at the beginning of the email, so that if your recipient is in a hurry he / she can quickly see the important points.

Everything you say must add value for the recipient so that they will read each detail rather than quickly scroll down to the end. However, you should also allow for the fact that they might scroll your email. Make it easy for them to do so by using:

• bullets

• bold to highlight important words or requests

• white space to separate items

When you have finished writing an email, ensure that it can be understood quickly and cannot be misinterpreted.

14.4 Use short sentences and choose the best grammatical subject

A lot of research has shown that when native English speakers read, their eyes tend to focus at the beginning and end of the sentence, whereas the middle part of the sentence tends to be read more quickly.

The way we read today is also very different from the way we read until the mid 1990s. The Internet encourages us to read very quickly – this is known as browsing, scanning or skimming. Because we want information fast in order to help us decide whether to respond to an email and what action to take, we tend not to read every word. Instead we skip from word to word, sentence to sentence, and paragraph to paragraph until we find information that we consider useful or important. If we don't find anything of value, we stop reading.

Essentially, you need to

- select the most important item to put as the grammatical subject of the sentence
- put the verb and object as close as possible to the subject
- limit yourself to a sentence with two parts – so that there is only a beginning part and an end part. If you have three parts (or more) the middle parts will be read with less attention

The sentence below is 64 words long. It is not too difficult to read.

> I am the sales manager at ABC, a company that produces engine parts for small aircraft and one of your clients, Stavros Panageas, kindly gave my your name as he thought you might be interested in receiving some documentation about the kinds of engine parts that would be suitable for your aircraft and which we believe would cost significantly less than your current supplier.

Nevertheless it requires more effort than this alternative:

> Your name was given to me by Stavros Panageas of XYZ. I am the sales manager at ABC, a company that produces engine parts for small aircraft. Mr Panageas thought you might be interested in receiving some documentation about the kinds of engine parts that would be suitable for your aircraft. We believe these parts would cost you significantly less than your current supplier.

The two texts above are the same length, but the information in the second text is much easier to absorb.

14.4 Use short sentences and choose the best grammatical subject (cont.)

Having short sentences:

- helps your recipients locate the key information in your sentence with the minimal mental or visual effort
- makes it much easier for you to change the order of the information. For example, you could switch the order of the first two sentences
- delete parts of your email or add parts to it
- makes it easier for the recipient to insert comments directly into the body of the email

The following are two versions of an email from a French student who wishes to do an internship in a company.

ORIGINAL VERSION (OV)

Dear Alyona Andropov,

I am Melanie Duchenne, the French student who Veronika Durov told you about few days ago.

Firstly, I would like to thank you for the opportunity you afford me to spend with your staff a short period, which would be extremely useful for me in order to obtain the master degree.

I have been adviced by Veronika to communicate to you my preference as soon as possible, and I beg your pardon for not having done it earlier, due to familiar problems. Then, if possible, the best option for me would be a two-months period, from the beginning of june to the end of july. Waiting for your reply, I wish to thank you in advance for your kindness.

best regards,

Melanie Duchenne

14.4 Use short sentences and choose the best grammatical subject (cont.)

REVISED VERSION (RV)

Dear Alyona Andropov,

I am the French student who Veronika Durov told you about.

Firstly, I would like to thank very much you for the opportunity to work with your team.If possible, the best option for me would be June 1 – July 31.

I apologize for not letting you know the dates sooner.

Best regards,

Melanie Duchenne

The RV is much more concise and precise. All non-essential information (from the recipient's point of view) has been removed. Reducing the amount of text reduces the number of mistakes. Below are the mistakes in the OV, with the correct version on the right.

few days ago = a few days ago

the opportunity you afford me = the opportunity you are giving me

obtain the master degree = to get my Master's [degree]

I have been adviced = I have been advised by Holger or Holger advised me

familiar problems = family problems

a two-months period = a two-month period

beginning of june to the end of july = beginning of June to the end of July

All of the above mistakes have been removed, simply by reducing the amount of text.

14.5   Choose the shortest phrase

Below are some examples of words and phrases that are typically used to introduce new concepts or link sentences together. For example, the following phrases could all be replaced with *because*:

because of the fact that

due to the fact that

as a consequence of

in the light of the fact that

in view of the fact that

And the following could be replaced with either *although* or *even though*:

in spite of the fact that

regardless of the fact that

However, do not confuse conciseness with brevity (i.e. using the minimum number of words). Brevity may have two major disadvantages:

- lack of precision and clarity
- it may sound rude and suggest that you couldn't find the time to make yourself polite and clear

14.6 Don't experiment with your English, instead copy / adapt the English of the sender

The less you write, the less chance you have of making mistakes with your English. Imagine that someone finishes their email to you in the following way.

> ... in the first quarter of next year.
>
> So in the meantime, Happy Christmas!
>
> Best regards
>
> Pete

In your reply it makes no sense to write something like this:

> Let me express my warmest wishes to you and your family for a very happy Christmas and a New Year full of both personal and professional gratifications.
>
> Best regards
>
> Raul

Raul's Christmas greeting is four times longer than Pete's Christmas greeting – the potential for making mistakes in English is thus substantially higher. Raul's greeting has two problems. It is:

- extremely formal compared to the rest of the email and thus sounds a little out of place
- probably a literal translation of what Raul would have said if he had been writing in his own language—however, the last part of the sentence (*full of both personal and professional gratifications*) does not exist in English (a Google search does not give any hits)

The best solution for limiting the number of mistakes you could potentially make is to repeat the other person's greeting:

> Happy Christmas to you too!

14.6   Don't experiment with your English, instead copy / adapt the English of the sender (cont.)

Copying the phrase of the sender and / or adding [*to you*] *too* is a good tactic for repeating a greeting, as the following examples highlight (the sender's greeting is on the left, the recipient's reply on the right):

I hope you have a great weekend.	I hope you have a great weekend too.
Have a great weekend.	You too.
Enjoy your holiday.	I hope you enjoy your holiday too.

Not all phrases can be replied to simply by adding *too*. For example, if the sender writes *See you next week at the meeting* you cannot reply with *See you next week too*. Instead, you could write: *I am looking forward to seeing you at the meeting*.

14.7 When using pronouns ensure that it is 100 % clear to the recipient what noun the pronoun refers to

A common problem in emails is the use of a pronoun (e.g. *it, them, her, which, one*) that could refer to more than one noun. The sentences in the first column below have been disambiguated in the third column.

AMBIGUOUS SENTENCE	REASON FOR AMBIGUITY	POSSIBLE DISAMBIGUATION
Thank you for your email and the attachment *which* I have forwarded to my colleagues.	*which* – email? attachment? or both?	Thank you for your email. I have forwarded the *attachment* to my colleagues.
To download the user manual, you will need a user name and password. If you don't have *one*, then please contact …	*one* – user name? password? or both?	To download the user manual, you will need a user name and password. If you don't have a *password*, then please contact …
Yesterday I spoke to Pete Jones, and on Tuesday I saw Jo Smith and one of his new staff, Vu Quach. If you want you can write to *them* directly. You will find their emails on the website.	*them* – Smith and Vu? all three people (Jones, Smith and Vu)?	If you want you can write to *all three of them* directly.
After a new employee has been assigned a team leader, *he / she* shall …	*he / she* – the new employee? the team leader?	After a new employee has been assigned a team leader, the *employee* shall …

As highlighted in the third column you can avoid ambiguity if you replace pronouns with the nouns that they refer to.

14.8 Avoid ambiguity

In an email you should make sure that you give it the same attention as any other important written document by making it 100 % clear and unambiguous. If you don't, it can be annoying for the recipient, who is often forced to ask for clarifications.

Ambiguity arises when a phrase can be interpreted in more than one way, as highlighted by these examples:

Our division is looking for candidates who speak English, Spanish and Chinese. Is the division looking for three different candidates (one for each language), or one candidate who can speak all three languages?

Each subscriber to a journal in Europe must pay an additional $10. Is the journal a European journal, or do the subscribers live in Europe?

You can't do that. Does *can't* mean that it would be impossible for you to do that, or that you don't have permission to do that?

14.9 Ensure that recipients in different time zones will interpret dates and times correctly

Businesses work in an international environment over many time zones. In the sentence below, it is not clear exactly when the server will and will not be available.

For maintenance reasons, the server will be not available tomorrow for all the day.

The problem words are *tomorrow* and *for all the day*. As I write this section I am in Italy and it is 17.00 local time. In Australia it is already *tomorrow*. What does *day* mean? Is it my day in Italy or my colleague's day in Australia?

Thus you need to be much more specific:

The server will not be available from 09.00 (London time) until 18.00 on Saturday 17 October.

To avoid misunderstandings due to differences between the ways various people write dates, I suggest you always write dates as follows:

12 March 2024

So: number of the day, then month as a word, then the year. If you write 12.03.2024, then this could be interpreted as 3 December or 12 March. Some people also write: March 12, 2024, but I find this less clear as the two numbers are together and a comma is also required.

14.10 Always check your spelling 'manually': don't rely on automatic spell checkers

Automatic spell checkers check the spelling of a document by searching for the existence of a word in their dictionary. This means that certain spelling mistakes will not be found, simply because the misspelled word also exists. For example,

> I got an email *form* Michel today.

Clearly, the spell checker does not look at the surrounding words to decide whether a particular word is spelt correctly. So it will not highlight that *form* should be *from*. Unfortunately, there are a number of less commonly used words in English which in emails are often typos (i.e. misspelled forms). For example, *whit* instead of *with*, *nay / any, sue / use* – in each case both words are in the dictionary.

So always double check your spelling. If you don't you may:

- create a bad impression – bad spelling is often equated with ignorance or someone who doesn't take care with their work

- confuse the reader – especially someone whose English is not as good as yours, who may think the word you have misspelled is actually spelt correctly

15 ABBREVIATIONS, ACRONYMS AND SMILEYS

This chapter is simply designed to be fun. It outlines many of the typical abbreviations used when writing very informal emails and text messages, and when chatting.

Note that such abbreviations, acronyms and phrases should NOT be used in a professional context.

15.1 Numbers

Numbers occur quite frequently in the abbreviations used in chatlines, emails and text messages. Due to the bizarre spelling system of English, numbers can be used in many different ways. For example the sound in 8 / eit / can have many different spellings: *eight, ate, ait* (as in *wait*).

1 / won /

1ce	Once
every1	Everyone
hag1	have a good one
ne1	Anyone
no1	no one
som1	Someone
sum1	Someone

2 / tu /

2	To
2	too
2b	to be
2day	Today
2moro	tomorrow
2nite	Tonight
f2f	face to face
f2t	free to talk
g2g	got to go
hrt2hrt	heart to heart

A. Wallwork, *Email and Commercial Correspondence*,
Guides to Professional English, DOI 10.1007/978-1-4939-0635-2_15,
© Springer Science+Business Media New York 2014

15.1 Numbers (cont.)

im2gud4u	I'm too good for you
se2e	smiling ear to ear
tlk2ul8r	talk to you later
wan2	want to
wan2tlk	want to talk?

3 / thri / or / fri /

ru32nite	are you free tonight?

4 / for /

4ever	forever
4yeo	for your eyes only
b@thpics4	be at the pictures [cinema] for 8 pm
b4	before
b4n	bye for now
j4g	just for grins
plz 4gv me	please forgive me

8 / eit /

bhl8	be home late
cub l8r	call you back later
cul8er	see you later
d8	date
gr8	great
h8	hate
l8	late
l8er	later
m8	mate
rungl8	running late
w8	wait
w84m	wait for me
w8in4u	waiting for you

15.2 Acronyms

A quick way of writing is to use acronyms, where each letter stands for a word.

a / s / l?	age / sex / location?
adn	any day now
afaict	as far as I can tell
afaik	as far as I know
afair	as far as I remember
afk	away from keyboard
aiui	as I understand it
aka	also known as
asap	as soon as possible
atb	all the best
ayor	at your own risk
bak	back at the keyboard
bbiab	be back in a bit
bbiaf	be back in a few (minutes)
bbl	be back late(r)
bbn	bye bye now
bbs	be back soon
bf	boy friend
bfn	bye for now
bg	big grin
botec	back-of-the-envelope calculation
brb	be right back
bsf	but seriously folks
bta	but then again …
btdt	been there done that
btw	by the way
bwl	bursting with laughter
cid	consider it done
clm	career limiting move
cm	call me
crb	come right back
cul	see you later
cyo	see you online
dak	dead at keyboard
diku	do I know you?
dos	dozing off soon
dqmot	don't quote me on this

15.2 Acronyms (cont.)

dtrt	do the right thing
dwb	don't write back
dwisnwid	do what I say not what I do
eiok	everything is OK
emfbi	excuse me for butting in
emp	extremely miserable person
eom	end of message
fc	fingers crossed
fish	first in, still here
fmtyewtk	far more than you ever wanted to know
foaf	friend of a friend
foc	free of charge
focl	falling off the chair laughing
fotflol	falling on the floor, laughing out loud
ftbomh	from the bottom of my heart
fud	fear, uncertainty, and doubt
fwiw	for what it's worth
fya	for your amusement
fye	for your entertainment
fyi	for your information
g	grin
ga	go ahead
gal	get a life
gd&r	grinning, ducking, and running
gf	girl friend
gfn	gone for now
gl	good luck
gmab	give me a break
gmta	great minds think alike
gtg	got to go
gtp	get the point / picture?
gtrm	going to read mail
gtsy	glad to see you
gw	good work
h&k	hug and kiss
hagn	have a good night
hand	have a nice day
hhis	hanging head in shame
hhoj	ha ha only joking

15.2 Acronyms (cont.)

hig	how's it going?
hitwth	hate it when that happens
ht	hi there
hth	hope this helps
hwru	how are you?
hyt	hey you there
iac	in any case
iae	in any event
ianal	I am not a lawyer (but)
idk	I don't know
iha	I hate acronyms
ihu	I hear you
iirc	if I recall / remember / recollect correctly
imho	in my humble opinion
imnsho	in my not so humble opinion
imo	in my opinion
ims	I am sorry
iow	in other words
irl	in real life
istm	it seems to me
iukwim	if you know what I mean
iwbni	it would be nice if
iyd	in your dreams
iykqim	if you know what I mean
iyswim	if you see what I mean
jam	just a minute
jic	just in case
jk	just kidding
jmo	just my opinion
jtlyk	just to let you know
k	okay
kiss	keep it simple stupid
kit	keep in touch
kwim	know what I mean
ld	later, dude
ldr	long-distance relationship
lho	laughing head off
lmso	laughing my socks off
lol	laughing out loud

15.2 Acronyms (cont.)

ltm	laugh to myself
ltns	long time no see
mcibtu	my computer is better than yours
mhdc	my hard disk crashed
mtf	more to follow
mtfbwu	may the force be with you
myob	mind your own business
nehw	no enjoyment, hard work
netma	nobody ever tells me anything
nfw	no feasible way
nhoh	never heard of him / her
np or n / p	no problem
nrn	no response necessary
nvm	never mind
nw	no way
o	over to you
obe	overtaken by events
obtw	oh, by the way
omg	oh, my god
oo	over and out
ot	off topic
otoh	on the other hand
ott	over the top
ottomh	off the top of my head
pcm	please call me
pmfjib	pardon me for jumping in but …
rehi	hello again
rl	real life
rotfl	rolling on the floor laughing
rt	real time
rtm	read the manual
sohf	sense of humor failure
sol	smiling out loud
somy	sick of me yet?
sot	short on time
sotmg	short on time must go
stw	search the web
sys	see you soon

15.2 Acronyms (cont.)

ta	thanks again
tafn	that's all for now
tanstaafl	there ain't no such thing as a free lunch
tcoy	take care of yourself
tgif	thank god it's Friday
tia	thanks in advance
tilii	tell it like it is
timtowtdi	there is more than one way to do it
tmb	text me back
tmi	too much information
toy	thinking of you
tptb	the powers that be
ttbomk	to the best of my knowledge
ttfn	ta-ta for now
ttt	thought that, too
ttyl	talk to you later
tu	thank you
tvm	thanks very much
ty	thank you
tyvm	thank you very much
uw	you're welcome
vbg	very big grin
wayd	what are you doing?
wb	welcome back
wbs	write back soon
wdalyic	who died and left you in charge?
wfm	works for me
wibni	wouldn't it be nice if
wrt	with regard to / with respect to
wt?	what / who the …?
wtg	way to go!
wu	what's up?
wud?	what you doing?
wuf?	where are you from?
wysiwyg	what you see is what you get
yattd	yet another thing to do
ybs	you'll be sorry
yw	you're welcome

15.3 How sounds of letters are used

Letters have always been used in English instead of words. The most commonly used until the advent of the internet was probably IOU which stands for *I owe you* to indicate that you owe someone money, for example *IOU $10* means *I owe you ten dollars.*

b	be
bcnu	be seeing you
c	see
cya	see ya
how ru	how are you
ic	I see
ilq	I like you
oic	oh, I see
qt	cutie
r u there?	are you there?
r	are
ru cmng	are you coming?
ru	are you?
ruok	are you ok?
sup	what's up?
thanq	thank you
tq	thank you
u	you
uok	you ok?
ur	you are
wru	where are you?
y	why

15.4   Use of symbols

In the examples below & stands for *and*, and @ for *at*, even in the middle of words.

l&n	landing
pl&	planned
po$bl	possible
s^	what's up?
th@	that
ura*	you are a star
x	kiss
xoxox	hugs and kisses
cu@	see you at

15.5 Contractions

The words below have been around for decades as part of the spoken language and also in rock, rap and blues lyrics. They are contractions of two or more words, which imitate the sound of English spoken quickly. In the table below, the first column is the contraction, the second the full form, and the third an example.

ain't	has not, am not	You ain't seen nothing yet.
betchu	I bet you	I betchu $100 that I am right.
betta	I had better	Betta go now.
coulda	could have	Coulda told you that myself.
cuppa	a cup of (tea)	Gonna have a cuppa.
dunno	I don't know	How much does it cost? Dunno.
gimme	give me	Gimme your email.
gonna	I am going to	Gonna tell all my friends.
gotta ...?	have you got ..?	Gotta minute?
gotta	I have got to	Gotta go now.
hiya	hi there	Hiya, how are you doing?
izzy	is he	Izzy someone special for you?
kinda	kind of	I kinda like it.
lemme	let me	Lemme get this clear.
lotta	a lot of	A whole lotta love.
mighta	might have	Mighta told me u were married.
outta	out of	Gotta get outta here.
shaddup / shadap	shut up	Shaddup will you?
shoulda	should have	Shoulda seen his face when I told him.
sorta	sort of	It's sorta like a dream.
soundsa	it sounds like a	Soundsa a good idea.
sup, wazzup	what's up	Sup mate?
wanna	want to	Wanna go out tonight?
watcha, wotcha	what are you, what do you	Wotcha gonna do about it?
wouldna	would not	I wouldna wanna do that again.

15.6   Short forms

For centuries English has been a language which has liked to simplify itself by reducing the length of long words. Some have been with us for so long that we no longer even notice that originally they were much longer, for example, *fax – facsimile*.

ad	Advertisement
add	address
brill	brilliant
coll	college
comp	computer
convo	conversation
cred	credit (on mobile phone), credibility
def	definitely
fav	favorite
info	information
min	minute
mob	mobile
mos def	most definitely
pic, pik	picture, photo
prob	problem
sec	second
tel	telephone
typo	typography mistake
uni	university

15.7 Abbreviations

An abbreviation just uses some letters (generally just the consonants) of the original word. It thus differs from a short form (15.6) which only uses the initial letters (both consonants and vowels) of a word.

b'day	birthday
bk	break
cfm	confirm
ctr	center
dnr	dinner
frm	from
grt	great
lsr	loser
lv	love
msg	message
n	and
nxt	next
pls / plz	please
ppl	people
rgds	regards
smt / smthg	something
spk	speak
thx, tnx, tx	thanks
txt	text
w / o	without
wknd	weekend
wrk	work
yr	your

15.8 Alternative spellings

Text messaging and chatlines have had a huge impact on the way English words are now being spelt in informal contexts. The spellings tend to be much more phonetical, i.e. to reflect more closely the way the words are pronounced:

flirtz, ladz, loadz	flirts, lads, loads (*z* = *s* plural and *s* third person singular)
gettin, interestin	getting, interesting (final *g* cut)

Sometimes the way a word is spelt will reflect the part of the English-speaking country where someone is from:

wid	with
wit	with
wiv	with

afta	after
alrite	alright
ansa	answer
av	have
'avin'	having
bin	been
bk	back
bout	about
coz, cuz	because
d, da or de	the
dem	them
dere	there
doin	doing
dun	done
eva	ever
ez	easy
fankz	thanks
fella	fellow
footie	football
gd	good
gudluk	good luck
hun	honey
iv	I've
jus	just
kann	can

15.8 Alternative spellings (cont.)

keul, kool	cool
laf	laugh
lata	later
lurve	love
luv	love
mite	might
n	and
nah	no
ne	any
nething	anything
no	know
nufn	nothing
'ome	home
ov	of
prolly	probably
recked	wrecked
rite	right
samink	something
sed	said
shopn	shopping
shud	should
skool	school
soz	sorry
spk	speak
spose	suppose
stewpid	stupid
ter	to
tho	though
uz	us
wel	well
wen	when
wi	with
wid	with
wit	with
wiv	with
wot	what
wud	would
wudnt	wouldn't

15.8   Alternative spellings (cont.)

wus, wuz	was
xlnt	excellent
yas	you
yer	yes

15.9 Smileys

According to Wikipedia: The smiley was first introduced to popular culture as part of a promotion by New York radio station WMCA beginning in 1962. Listeners who answered their phone "WMCA Good Guys!" were rewarded with a "WMCA good guys" sweatshirt that incorporated a happy face into its design.

Smileys are also known as emoticons.

To see a very comprehensive list of Japanese emoticons: japaneseemoticons.net /

The table below shows how creative people have been in using the characters of the keyboard to create 'pictures'.

Amazed	:<>
Angel	O:-)
Angry	:-II
Bald intellectual	(:I
Baseball	d:-)
Bawling	:~-(
Beaver	:=
Been on the town all night.	#-)
Big Face	(:-)
Big Hug	(((H)))
Big Kiss	:-X
Blabber Mouth	(:-D
Black Eye	?-(
Blockhead	:-]
Bow Tie-Wearing	:-}X
Brain Dead	%-6
Bucktoothed	:-(=)
Can't believe it	:-C
Cat	}:-X
Censored	(:-#
Chef	C = :-)
Chin up	;-(
Clown	*<):o)
Confused	:-S
Cross-Eyed	H-)
Crying	:'-(
Crying	:'-(

15.9   Smileys (cont.)

Crying	<:'-(
Crying softly	:*(
Curly hair	&:-)
Cursing	:-@!
Cyclops	O-)
Devilish	>:->
Disappointed	:-e
Disgusted	:-|
Dog	:3-]
Doing nothing	o-&-<
Double chin	<:-))
Drinking every night	:*)
Drooling	:-).....
Drooling	:-P
Drunk	:#)
Drunken smile	%-}
Dunce	<:-I
Embarrassed	:")
Evil grin	>:)
Foot in Mouth	:-!
Forked Tongue	:-W
Frenchman with a beret	/:-)
Frog	8)
Getting tired, asleep	|-)
Got a cold	:-~)
Had a fight and lost	% + {
Happy	:-)
Homer Simpson	(_8(|)
Hungover with headache	%*@:-(
Hungry	:0
I won the lottery	$-)
In a hurry	O-S-<
Just back from hairdresser	@:-}
Kiss	:-*
Laughing	:-D
Left handed	<(-:
Lost contact lenses	|-(
Mad	X-(

15.9   Smileys (cont.)

Makes me cry	&-l
Makes me sick	:-(*)
Makes no sense	:-S
Mickey Mouse	8(:-)
Moustache	:-{
My lips are sealed	:-#
Oh my god!	8-O
Orangutan	:=)
Penguin	8>
Persian cat	:<
Pig	:8)
Pinnochio	:---)
Pirate	P-(
Pope	+<:-)
Priest	+:-)
Prizefighter	;~[
Punk	=:-)
Really angry	<:-[
Robot	[:]
Rose	@ >--;--
Rudolph the red nose reindeer	3:*>
Sad	:-(
Sad and lonely	:-<
Sad punk	= :-(
Said with a smile	:-d
Santa Claus, Father Christmas	*<|:-)>
Sarcastic smile	<:->
Skeptical	:-/
Screaming	:-@
Shouting	:-V
Side splitting laughter	:-D
Single Hair	~:-P
Skater	o[-<]:
Smirking	:^)
Smoking	:)-~
Smoking a cigarette	:-i
Smoking a pipe	:-?
Someone gave me a black eye	%-(

15.9 Smileys (cont.)

Sunglasses face	8-)
Surprised / shocked	:-O
Talkative	:-0
Tongue in cheek	:-P
Tongue tied	:-&
Turkey	<:>= =
Undecided	:-\
Vampire	:-[
Very Happy	:-))
Very Unhappy	(:-(
Very Very Happy	:-)))
Walrus	:-<
Wearing glasses	B-)
Wearing lipstick	:-{}
Wearing sunglasses	]-I
Whistling	:-"
Winking	;-)
Wizard	8 <:-)
Wizard with Wand	- = #:-)\
Yawning	I-O
Yelling	:-(0)
Yummy	:-9

16 USEFUL PHRASES: GENERIC

Every language has certain phrases that cannot be translated literally into another language. A high percentage of the content of emails is made up of such standard phrases. You need to be very aware of what these standard phrases are, and what their equivalents are in English.

In addition to the useful phrases listed in this chapter and in Chapter 17, you could create your own personal collection of useful phrases, which you can cut and paste from emails written by native English speakers (which hopefully will be correct!).

If you make literal translations into English, the result may sound strange or even comical and thus sound unprofessional. Here are some examples of expressions whose literal translation sounds very strange and inappropriate in English.:

GERMAN	JAPANESE	RUSSIAN
Beautiful greetings.	To omit the greetings.	Healthy.
I feel pleasure for myself from you to hear.	Thank you for supporting us always.	Wish a success
Say you a greeting to your wife.	Please kindly look after this.	Calm night.

Using standard phrases enables you to be sure that at least the beginnings and ends of your emails are correct! Then in the body of the email it is advisable not to experiment too much with your English.

Note that the use of English varies from one English-speaking country to another. If you are from India, Pakistan, Bangladesh and other countries (e.g. in Africa) with strong historical ties to England, then your standard English usage may be considerably more formal than, for instance, in the UK and the US.

People in the West tend to be less deferential to their superiors and use considerably fewer salutations at the end of an email. An expression such as *Sincerely yours* that might be considered perfectly acceptable by Indian speakers of English, sounds much too formal or even rather archaic to someone in the UK or US, where even *Yours sincerely* tends to be reserved for very formal letters. A much more typical salutation is *Best regards*, which works both in formal and more neutral situations.

A. Wallwork, *Email and Commercial Correspondence,*
Guides to Professional English, DOI 10.1007/978-1-4939-0635-2_16,
© Springer Science+Business Media New York 2014

16.1   Initial salutation

Standard

Dear James

Dear James Bond

Dear Mr Bond (formal) [NB Mr – male, Ms – female]

Someone you know well

Hi!

Hi there

Hope you are keeping well.

Hope all is well.

Group / team

Dear all

Hi all

Hi everybody

To all xxx members and xxx development group

FAO xxx Support (= for the attention of)

To xxx Support

A specific role

Dear Human Resources Manager

Dear Sales Manager

For the kind attention of the Marketing Manager

Someone / some people whose names or job positions you don't know

Hi

Hello

Good morning

To whom it may concern

Dear Sir / Madam

16.2   Introducing yourself to people who don't know you

Colleague in the same company

Good morning. I work with Hans in business development. I'm responsible for …

I just wanted to take this opportunity to introduce myself … You will be reporting to me for …

Hi Silvia, It is nice to 'meet you'. I'm Nicola in the accounts department.

Potential client, explaining your company and position

My name is Megawati Hok, I look after the *position* for *company* in Jakarta.

ABC is a large IT company operating out of *place*. I am the marketing manager and I would appreciate an opportunity to show you …

I am writing to inquire whether you might interested in buying …

I wish to offer you my services for the purchase / sale of … in *place*.

I have had over … years' experience in this business.

I am the agent for … and I wish to …

Someone you have met only once

You may remember we met last year …

You may recall being in contact last year …

We met at the Nuremberg Trade Fair last August, where you gave me your business card.

Checking that the recipient is the right contact person

I don't know if you are the right person to ask but …

If you are not the right contact, please could you let me know who I should contact.

If there is someone who is more suitable for me to contact, maybe you could forward this email to them.

Saying where you found the recipient's name

We were given your address by a mutual acquaintance, Ms Anna Southern.

Your name was given to me by … I was just wondering whether …

I found your email address on the web, and am writing to you in the hope that you may be able to help me.

Your address was given to us by …

I saw your company's stand at the Bahrain Expo and I would like to have an opportunity …

I found your name through various contacts on LinkedIn.

I understand from Hartono Swee King that you wish to have an agent in this market.

Flores Bautista suggested I contact you.

Saying how you know about the recipient's company

Your firm has been recommended to us by ...

We have seen your advertisement in ...

We saw your stand at the ... Fair / Exhibition.

16.3   Making an inquiry (first contact)

Beginning

I am writing to you because ...

I was wondering if by any chance you ...

I wonder if you might be able to help me.

I would be extremely grateful if you could ...

I have a couple of simple requests:

Could you please tell me ...

I would like to know ...

Would / Could you please send us ...

Could you possibly send me ...

I have some questions about ...

Please can you help us with the following:

My questions are:

Ending

We look forward to hearing from you soon.

Any information you could give us would be very much appreciated.

I would be grateful for any further information you may be able to give me about ...

We thank you in advance for your prompt reply.

Thanks in advance.

16.4   Responding to an inquiry

Thanking

Thank you for your interest in …

Thank you for your letter of the … informing us that …

Thank you for your letter of the … and have much pleasure in replying to your various questions as follows:

Saying what you have done or will do

In reply to your letter (inquiry) of the …, we wish to inform you that …

We have passed on your enquiry to our …

Regarding your queries about …

In response to your questions:

Here is the information you requested:

As requested, I am sending you …

Below you will find the responses to your points re …

Here are the answers to your questions point by point:

As agreed please find attached …

Asking for details

Before we can proceed with your order, we need further details re the following:

Before we can do anything, we need …

Please advise us ASAP how you wish to proceed.

Adding details

Please note that …

We would like to point out that …

As far as we know …

We wish to add that …

I'd also like to take this opportunity to bring to your attention …

May I take this opportunity to …

Telling client they can ask for further info

Please feel free to e-mail, fax, or call if you have any questions.

If you need any further details do not hesitate to contact me.

Should you have any questions please let us know.

Please do not hesitate to contact us should you need any further information.

Please contact me if you need any further clarifications.

For further information about ... please email me or ring me on: 0039 ...

Any questions, please ask.

Hope this is OK. Please contact *name* if you need any further details.

Ending

Please let me know if this helps.

We hope to be able to give you a definite answer soon.

Once again, thank you for your inquiry.

Many thanks for your continuing interest in ...

16.5 Making reference to previous correspondence

Reference to previous mail / phone call / conversation

Re your inquiry ...

Regarding ...

In relation to ...

With reference to ...

Further to our conversation of yesterday ...

Further to our recent meeting ...

Following your contact with *name*, who is responsible for your area, we are very pleased to ...

In reply to your email of 12 April, I would like to inform you that ...

As I informed you yesterday ...

Reference to the main body of their email / letter / fax

In accordance with your instructions, we ...

We note your remarks re ...

We understand that ...

As requested I am sending you ...

Reference to crossed email / letter / fax

Your letter dated 17 August crossed ours of the same date.

Sorry, but it looks as if our emails got crossed. In any case here's the information you required:

16.6   Making requests to people who already know you

Requesting help / advice

Please could you …

Would you have any suggestions on how to …

It would be very helpful for me if I could pick your brains on …

I would like to ask your advice about …

I'd value your input on whether …

I would really appreciate your help in …

Have you any thoughts on this?

Showing awareness that you are taking up recipient's time

I realize you must be very busy at the moment but if you could spare a moment I would be most grateful.

If it wouldn't take up too much of your time, I would be very grateful if you could …

Clearly, I don't want to take up too much of your time but if you could …

Obviously, I don't expect you to …. but any help you could give me would be much appreciated.

I know it's Friday afternoon and you are probably pushed for time, but do you think you could possibly send me the info on …

Specifying what you have attached and what recipients are expected to do

I have attached our terms and conditions for your review.

Attached is the agenda – if you could just quickly look through it and give me your thoughts.

First off, could I please ask you to … Secondly, If you have any suggestions regarding x then I would be very grateful.

I'd welcome any suggestions that you may have regarding …

In any case it would be great if you could let me know what changes are to be made.

Concluding

I would be grateful to hear your comments.

I look forward to hearing your thoughts.

As I said, it would be great to hear your opinion on this.

Do let me know if there is any more information you need from me.

Full details are enclosed for your reference.

Please let me know your decision on this matter.

Your early reply would be appreciated.

If you think this would be possible for you, then I will get the OK from my management.

Please let me know how you feel we should proceed.

16.7   Making announcements and giving instructions

Announcing decisions

To all concerned, we are now moving forward on ….

The following issues and minutes are based on my meeting with …

We will resume our weekly conference call on …

A weekly status report will be produced…The report will show …

Telling recipient(s) how you want them to proceed

Could you please …

Please have a look at …

If you could …

I would be grateful if you could …

Please review the attached draft project plan for … and please make any additions / suggestions by Monday.

Please let me know if there are any issues or concerns with my requests.

Asking for a reply

I look forward to hearing from you …

... in the near future.

... soon.

... before the end of the week.

Please could you get back to me.

... by the end of today.

... this morning.

... ASAP.

I hope you can reply this morning so I can then get things moving before leaving tonight.

We would appreciate a reply as soon as possible.

Looking forward to your reply.

I would appreciate your immediate attention to this matter.

Appreciate your early reply.

16.8 Replies to requests from people who you know

Friendly beginning

Great to hear from you. I have had a chat with ... and ...

Nice to hear from you again.

Neutral beginning

Thanks for your mail. It will take me a while to find all the answers you need but I should be able to get back to you early next week.

Re your request. I'll look into it and send you a reply by the end of the week.

Accepting

No problem. I'll get back to you as soon as ...

I'd be happy to help out with ...

I'd be happy to help.

In response to your other questions:

Explaining why you cannot help

I'm sorry but ...

I'd like to help but ...

Unfortunately …

At the moment I'm afraid it's just not possible.

Sorry, but I'm actually going on holiday tomorrow, so I'm afraid I won't be able to get back to you for a couple of weeks.

I would be happy to discuss the requirements with you but I'm off on leave for two weeks so I am afraid we'll have to postpone the discussion until I get back.

In the meantime if there is anything else you need clarifying then do drop me a line.

Asking for further details

Before we can do anything, I need a few more details.

I am very happy to help but first can you tell me whether …

Do you want me to …?

Would you like me to…?

Shall I …?

Do we need to …?

Let me know whether …

Telling recipient when you will reply

I'll get back to you before the end of the day.

I will get back to you on this by tomorrow evening.

I'm sorry but I won't be able to give you any response until …

I will contact you as soon as we have something more definite to tell you.

You will hear from us as soon as possible about the matter.

As soon as we are able to say anything definite, we will write to you again.

I will contact you when I return.

Will be in touch.

16.9   Replying on behalf of someone else

Explaining you are not the right person

I am not the person you need to contact. I will forward your request to *name* who is our head of Sales. If you wish you can contact him / her directly on:

Re your e-mail to our help desk, please note that Joe Bloggs handles all our advertising.

I'm marketing manager at *company*, and Jon Kennedy works at our help desk. So please refer to Jon for any question related to the help desk, whereas you can call me if you need information on …

I am not actually the right person to comment on … You would better talking to …

When someone has asked you to do something

Dominique Batteaux asked me to send you …

I have been instructed by Helena Gooley to …

We have been informed by …

When someone's request has been passed on to you

Thank you for sending us this file, which was passed to me by my colleague Leo Tolstoy. I have recently moved from our R&D division, where I was responsible for…

I have recently joined *company* as Sales Manager for your area, and Gordon Brown passed on your request to me.

Your message has been forwarded to me. I am the … I will be happy to …

16.10   Replying to someone who has just replied to you

Initial thanks

Thank you for getting back to me.

Thank you very much for your useful reply.

Thanks for the quick response!

Wow! That was quick.

Expressing your appreciation

Your comments are really useful – thank you so much.

I very much appreciate you taking the time to deal with this.

Concluding

Thanks again for all your help, We will keep you informed of the status and may need to get back to you for more details.

Once again, thanks very much.

16.11   Chasing

Soft approach

I was wondering if you had had time to look at my email dated 6 June (see below).

Did you get my last message sent on …?

I was wondering whether you had received my e-mail (sent on 2 June).

I would be grateful if you could let us have your answer to our letter of the … concerning …

I wonder if you could now give me some definite information?

Stronger approach

Could you let me have an answer as soon as possible to the question I raised in my email of last week (see message below).

This is the third time we have raised this problem.

I would like to remind you that I still have not received an answer to my question.

I am a little concerned that for a few weeks I haven't had any feedback from you on …

As I am sure you are aware, if we don't …. then we might end up … so we risk …

May we remind you that we are still awaiting your reply to our letter of the … regarding the … Your reply is urgently requested.

As the information requested in our letter of the … is now urgently required, your early reply will be greatly appreciated.

16.12   Responding to a chase

Apologizing

I'm sorry it has taken me so long to get back to you but …

We apologize for the delay in replying to your email dated …

My sincere apologies.

Asking for patience

Thanks for the reminder. I haven't forgotten, but it has been difficult for us to make time to meet and discuss it.

If you can bear with us, we will get round to it very soon.

Saying when you plan to take action

I will give it top priority as of tomorrow morning.

My aim is to get these authorized by close of business Friday.

16.13   Reporting on progress and updating

Beginning

Just a quick update on …

Just to let you know that …

This is to let you know that …

This is just a quick message to …

I'm writing to inform you that …

We are pleased to inform you that …

We are writing to inform you about …

This email is intended to inform you that …

Saying who you've contacted

I spoke to Jim yesterday and he said that …

I have spoken to Jim and he says that …

Saying what you've learned from others

Darren has just told me that …

We've been informed that …

Since my last email to you I've ascertained that …

Valerie mentioned that … She said … and wants to know if …

Abid said he had spoken to …

Saying what you've done

I've looked into the matter and have taken measures to …

At the last board meeting it was decided to …

Saying what you will do next

I will send you all the details re … in due course.

With regard to your email dated 10 March, I will talk to my colleagues and get back to you ASAP.

Reporting what you said you would do

I said that I would check on this and would get back to him.

I advised him that …

Asking confirmation if what you have done is acceptable

I hope that is OK – if not please raise with Mike.

Is that OK?

Stating what action you expect from recipient

Could you please call him to give him this information once you get in?

Will you check with her that this is OK?

Asking to be kept informed

Keep us informed of any developments.

Please let us know the outcome.

Please keep me posted.

16.14 Contact details

Asking for

Please let me know / have your full address so I can send you the offer for *product / service*.

Please could you tell me who I should contact regarding …

I am a bit confused about who to contact regarding this matter.

Giving

Please contact our help desk at:

The contact person is …

My contact details are as follows:

For future reference all correspondence regarding x should be forwarded to …

16.15   Invitations

Making

Are you free for lunch on Monday?

Are you able to join us on Friday? We're having dinner to discuss …

We're having a meeting on Thursday which I do hope you will be able to attend.

We would be delighted if you could join us on …

Accepting

22 April is fine with me.

I would love to come.

I'd be delighted to join you then.

I will be delighted to accept.

I accept with pleasure.

Declining

Sorry but I can't make it that day.

Sorry but I'll be on holiday then.

I'm afraid I have another engagement on 22 April.

Declining after initially accepting

Sorry but I am going to have to cancel lunch next week.

Due to family problems I am sorry to inform you that I am no longer able to attend the meeting on …

16.16   Making arrangements

Asking a meeting / phone call

Let's arrange to meet and we can discuss it further.

Can we a arrange a meeting on …?

Would it be possible for us to meet on …?

Could you let me know the best number to call you on? It's probably easier to have a quick chat rather than just emailing.

I would be grateful if you could contact me on this email address to arrange a possible meeting.

Would it be ok for you and I to have a 30 min call to talk about ... ? Maybe on Monday morning before the week gets going?

Requesting training, demos, presentations etc

We'd like to organize a training session for next week (one each day if possible excluding Monday), preferably Tuesday afternoon, Wed / Thu (morning or afternoon) and Friday morning.

When would you be free to come in and give us a presentation on your solution? Please let me know when you are available.

I'm e-mailing you as I'm planning to come to *place* next week. I thought this might be a good opportunity for me to give you a presentation of *product*.

Suggesting the time

How about Wednesday straight after lunch? Will try and call you Monday to confirm.

Would Wednesday after lunch suit you? I will contact you again on Monday to confirm.

Would 10 March suit you?

What / When would be a good time for you?

Asking for confirmation

Let me know a suitable time if you think this is a good idea, and if you have any ideas or thoughts.

Please could you confirm as soon as possible. If you are not available for the suggested times, please give me some other options for this week.

Please check your schedules to see if there are any conflicts with these dates.

Would you please confirm your attendance, and let me know if you have any further items which you would like to be added to the agenda of the meeting.

Your cooperation is requested in ensuring that the meeting is a success.

16.17   Fixing the time

Initial reply to request for meeting / phone call

I would love to discuss, let's schedule some time.

That's sounds like a great idea. Friday morning would be good for me.

I've just seen your mail now. I really don't think I'll have a chance to talk to you today as shortly I am going to be leaving the office.

I have tried getting through to you to arrange a time, but unfortunately have not been able to.

Unavailability at that time

Would love to meet – but not this week! I can manage November 17 or 17 if either of those suit.

I am afraid I won't be available either today or tomorrow. Would Thursday 11 March suit you? Either the morning or the afternoon would be fine for me. I'd be grateful if you could let me know as soon as possible so I can make the necessary arrangements.

Changing the time

Sorry, can't make the meeting at 13.00. Can we change it to 16.00? Let me know.

Re our meeting next week. I am afraid something has come up and I need to change the time. Would it be possible on Tuesday 13 at 16.00?

We were due to meet next Tuesday afternoon. Is there any chance I could move it until later in the week? Weds or Thurs perhaps? Please let me know your availability.

Can we fix it for another time?

Confirming the time

I look forward to seeing you on 30 November.

OK, Wednesday, March 10 at 11.00. I look forward to seeing you then.

16.18   Giving directions on how to reach your office

To get here, the best thing would be to take a taxi.

The easiest route is as follows:

The nearest airport is Pisa which is only a ten minute taxi drive from our office. If you are coming by train, Pisa Centrale station is only a five minute walk.

The nearest station is …. and we are only a fifteen minute walk from the station. Alternatively you could take a taxi and ask for ….

On your arrival, ask at the reception for *name*.

Here's the address again just in case:

16.19   Favors / giving help

Asking

Please could you …

I was wondering if by any chance you …

I wonder if you might be able to help me.

I would be extremely grateful if you could …

I found your email address on the web, and I am writing to you in the hope that you may be able to help me.

I realize you must be very busy at the moment but if you could spare a moment I would be most grateful.

Accepting

No problem. I'll get back to you as soon as …

I'd be happy to help.

Refusing

I'm sorry but …

I'd like to help but …

Unfortunately …

At the moment I'm afraid it's just not possible.

16.20   Thanking

For initial contact

Many thanks for this / your e-mail.

Thank you for visiting our website.

For quick response

Thank you for the quick response.

Thanks for getting back to me.

Thank you very much for your prompt response to my fax.

For feedback, suggestions, advice etc

Thank you for the e-mail. We appreciate your feedback, and will get back to you as soon as possible.

Suggestions are always welcome, thank you.

Thank you very much for your warm words of …

For help with work done together on site

Anyway, I just wanted to thank you for all the time you all have dedicated to me this week. It was really useful for me and I am already thinking about how I can put everything you told me into practice.

Thanks again and I hope we will have an opportunity to meet again in the not too distant future.

In advance

Thanks in advance.

Thanks for any help you can give me.

Thank you very much for your assistance.

I thank you in advance for your cooperation.

May I thank you for your help in this matter.

Ending the message

Many thanks for this.

Thanks once again for all your trouble.

Thanks for your assistance!

Thanks for your help in this matter.

Thank you for your help in solving this problem.

My apologies for bothering you, and thank you once again for your kind help.

Cheers.

16.21 Opinions

Expressing opinions

I think …

I am much in favor of …

As I see it …

In my opinion …

It is doubtful that we can …

Agreeing

OK.

Fine.

Right.

Yes, I agree.

I completely agree with you re …

Disagreeing

I'm not sure I agree with you about …

I think that we probably see things rather differently.

The main questions which seem to divide us are …

Making suggestions

The only solution might be for xxx to do yyy.

It seems to me that there is a simple way of putting this right.

I suggest that …

Maybe we could …

Asking for suggestions

If you have any suggestions to make re …, I would be glad to hear them.

What do you recommend?

Any suggestions?

What are the options?

What do you suggest?

16.22   Asking for and giving clarifications

Asking for clarification

I'm not sure what you mean by "xxx"

What exactly do you mean by "xxx"?

Sorry, what's an "xxx"?

Checking that you've understood

I'm assuming you mean …

Do you mean that …?

Giving clarification when reader tells you he / she didn't understand

What I meant by xxx is …

My point is that …

In other words …

So what I'm saying is …

Giving clarification when you realise reader didn't understand

Sorry, no what I meant was …

Sorry about the confusion, what I actually meant was …

Sorry I obviously didn't make myself clear.

Giving clarification when you realize later that you have not been clear

I've just reread the message I sent to you earlier and I think I need to clarify a few points.

Following my earlier email to you, I suddenly realized that I may not have made it clear exactly what I wanted. Basically I need …

Hoping you have been clear in your clarification

I hope this helps clarify the problems.

Let me know if there's still something that is not clear.

Replying when you have been given clarification

OK, understood.

OK, I'm clear now.

OK, but I'm still not clear about …

16.23   Apologizing

For not answering an e-mail

Sorry for the delay in …

   getting back to you.

   replying.

   sending you the information you requested.

Apologies for the late reply.

I have been out of the office this week.

I've been away for the last few days.

Sorry, but I have only just read your message now.

Sorry, but our server has been down, so we haven't been receiving any mails.

Sorry but we've been having emailing problems.

I apologise for the delay in responding but as Ms X has left the company, I think your original request got lost in the re-organisation.

Expressing regret in general

I am sorry to inform you that …

We're sorry that we're not able to provide you with the information you requested.

I'm sorry that I can't give you a more specific reply.

I am sorry for the inconvenience this may have caused you.

Please accept my apologies.

I'm really sorry about this.

For your e-mail not arriving

For some reason my last email had delivery problems. So here it is again just in case you didn't get it the first time round.

Please reply to the above address as our regular connection is down. Thanks very much.

For bothering someone again

Sorry to bother you again.

For sending a blank e-mail

Sorry I accidentally hit the send button.

For a misunderstanding

OK, I'm sorry – you are right. I misunderstood.

Sorry about that, we obviously had our wires crossed!

Sorry for the confusion.

For not being able to say goodbye after on-site meeting

I am really sorry that I missed saying goodbye to you yesterday.

Sorry that we didn't get the chance to say goodbye yesterday.

Repeating apology at end of mail

Again sorry for the delay.

Once again, apologies for any trouble this may have caused you.

Once again please let me apologize for what has happened, and I assure you that in the future we will do our best to avoid any similar problems.

16.24   Sending documents for approval

Asking for comments

Once you have reviewed the document, please forward it to …

Please let me know if see any need for additions or deletions.

Please let me have your feedback by Friday so I can send you a draft programme next week.

Could you please check these comments and let us know if you still have any issues with …

Please have a look at the enclosed report and let me know what you think.

Giving positive comments

I have now had a chance to look at your report, it looks very good.

I was really impressed with …

The only comments I have to make are …

Hope this helps.

Disguising negative comment by making a suggestion

While I like the idea of … I am not convinced that …

I'm not sure whether …

It might not be a bad idea to …

Have you thought about …?

It seems that …

Replying to comments

Many thanks for this. All points noted.

Yes, I see what you mean.

Thanks your comments were really helpful.

Fine, I'll see what I can do.

16.25   Sending and receiving attachments, faxes, emails

Telling recipient about your attachment

I'm attaching …

Please find attached …

Attached is / are.

Let me know if you have any problems opening the file.

As you will see from the attached copy …

Asking for confirmation of receipt

Please confirm / acknowledge receipt.

Please ack receipt.

I'd appreciate it if you could confirm your receipt via either fax or e-mail.

Please could you acknowledge receipt of this mail as we are not sure we have your correct address.

I would be grateful if you could confirm that you have received it.

Giving confirmation

I have just received your fax.

I confirm receipt of your fax.

This is just to confirm that I received your mail. I will get back to you in / on / at …

Telling sender you couldn't read the mails / attachments

Sorry I couldn't read your mail.

I received your mail, but I'm afraid I can't open the file.

When I try to open the file the system crashes.

Telling sender they forgot to do something

Thanks for your mail but I'm afraid you forgot to

   send the attachment

   include the prices

   tell me when you will be available

We received your message dated 25 September but you omitted to send details of …

Apologizing for forgetting to send attachment

Sorry, I just sent an email without the promised attachments.

Sending attachment again

Sorry about the problems. Here's the attachment again. Let me know if you still can't read it.

There must have been some problems on the network. I am attaching the document again. Please could you confirm that you have received it.

16.26   Fax transmission and scans

Problems in sending

I've tried your fax number several times but have been unable to get through.

Could you send your fax number again as I think I must have the wrong number.

Problems in receiving

Could you please send the fax / scan again as it was too faint to read.

We only received three pages of your six-page fax.

Could you send the last two pages again please.

I am afraid the scanned documents you sent were not legible.

Asking for confirmation of arrival

I sent our order today by fax. I hope you received it successfully.

Could you just confirm that you have received our fax.

16.27   Phone calls

Problems in contacting someone

I have been trying to contact you over the phone but with no luck …

I have left several messages with your secretary but …

Making reference to a phone call

With reference to our phone call of …

Re our phone call this morning.

Further to our telephone conversation, here are the details of what we require.

Following up a call

Thanks for ringing me yesterday.

It was good to speak to you this morning.

As I said / mentioned on the phone ...

I just wanted to check that I've got the details correctly.

Many thanks for your earlier call. As discussed, ...

16.28 Booking hotel rooms

I would like to book two single rooms with bath in the name of: xx and yy. Arriving 10 March, leaving 12 march. Both guests will be arriving at around 19.30 on March 10.

Please could you email me confirmation or fax me on: 0034 050 500 0007

16.29 Circular emails

Out of office message

I am on leave from Monday 07/08 to Wed 17/08. If you have any problems or queries please contact the IT office on x.1234.

I'm out of the office all day today but will get back to you tomorrow re any urgent messages.

I am out of the office. For urgent matters please call the Sales general number: 0208 ...

Change of address / phone etc

Please note that as of now my email address is:

This is to inform you that as of this coming September 10 our offices will be transferring to the address given below.

We are pleased to be able to inform you that we have moved our offices and warehouse to new and larger premises at the following address:

We are writing to inform you of our change of telephone number. As from July 10 the initial code on all our numbers will be 0171 rather than 061.

We have today moved to ...

Our showrooms have been transferred to ...

Announcing celebrations

To celebrate 10 years of business with you, we would like to offer you an exceptional discount of 20 % on all our product range.

To celebrate the 50th anniversary of the founding of our company we ...

Change in staff / positions

We are pleased to announce the appointment of our news sales manager, Ms Nelly Sparks. As from next week, all sales enquires should be addressed to her at:

We regret to announce that James Henpeck no longer works with us. His position will be taken by ...

I have recently joined *company* as Marketing Manager for Europe ...

16.30   Chit chat with colleague you know well

Beginning of email

I hope you are well and had a great weekend! How are things going in *place*?

How are things?

How did the holiday go?

How was the match? Did you win?

End of email

Hope you are enjoying the summer, and look forward to hearing from you.

Have a great weekend.

Enjoy your holiday

I look forward to hearing from you, wishing you a lovely week ahead!

16.31   Sending regards and wishes

Sending regards to third parties

Say hello to ...

Please send my regards to ...

Remember me to ...

Please convey my best wishes to ...

Sending regards on behalf of a third party

Mrs Southern sends her kindest regards.

Anna says hello.

Sending holiday wishes

Best wishes for the holidays and the new year from all of us here at *company*.

Have a great Thanksgiving!

Have a nice weekend.

Happy Easter to everyone.

May I wish you a …

I would like to take this opportunity to wish you a peaceful and prosperous New Year.

Please accept our sincere good wishes for Christmas and the New Year.

Health problems and condolences

I was really sorry to hear about …

I can imagine this must be a very difficult time for you …

If there is anything I can do to help, then don't hesitate to let me know.

We are glad to have news of … and to hear that he's making good progress. Please convey our best wishes and kindest regards to him.

We have learned with deep regret the sad news of the death of

I was deeply grieved to hear of the sudden death of …

I would like to express my heart-felt sympathy.

Please convey my sympathy also to his family.

16.32   Final salutation

Neutral

Kind regards

Best regards

Best wishes

Regards

With kind regards

With best wishes

Informal

All the best.

Have a nice weekend.

See you on Friday.

Hope to hear from you soon.

Speak to you soon.

Cheers.

Formal business letter

Yours faithfully

Yours sincerely

Sincerely yours

Yours truly

17 USEFUL PHRASES: COMMERCIAL

17.1 Making inquiries

Asking about products and services

We are interested in …

We are considering buying …

We understand from … that you produce …

We urgently need …

Do you offer a …?

Could you also let us know if you are prepared to …

Will you please let us know your prices for …

We would like to know whether you could supply …

Could you kindly give us a quotation for …

Please send us your price list for …

Could you send us further details of …

I would be grateful to receive samples of …

Please could you email us your catalogue along with further details of …

Asking for an estimate

Please submit an estimate for …

Could you send me a quote for the following:

Could you please quote your best price for the supply of …

Please quote for the following:

Informal inquiries about orders

What do I need to do to order a …?

I would like to know if I can order an xxx directly from you (our company is located in *place*).

I would like to know if it is still available and if so what I need to do to buy a copy.

I am looking for an xxx. Do you have one in stock?

A. Wallwork, *Email and Commercial Correspondence,*
Guides to Professional English, DOI 10.1007/978-1-4939-0635-2_17,
© Springer Science+Business Media New York 2014

Asking about pages on a website

I have looked for "name" directory on your website, but I couldn't find it. Could you please send me the complete path …

I am unable to find xx that you mentioned – is it on the website? All I can find is …

I am unable to download the catalog from your site.

Making follow up enquiry

Thank you for … Would it be possible for you to send me a bit more information on …?

Could you please describe what is included in the …

Could you please send me further details about your products along with some samples.

Could you email me your sales conditions.

What discount could you offer us on orders of over 10,000?

17.2   Replying to inquiries

Thanking

Thank you for your inquiry dated 6 June. We will be sending you under separate cover a …

Thanking you for your inquiry dated 27 August. We can supply *product* for immediate delivery …

In reply to your email dated 6 January, we have pleasure in submitting the following quotation:

Thank you for your order No. 123, for which we attach our official confirmation.

Replying positively to product inquiry

We are pleased to confirm your order No. 9678568 dated 22 April.

We have received your order of …

Yes, we do have *product* in stock.

We currently only stock x, but getting y within 3–4 days shouldn't be a problem.

Re your recent enquiry, we can offer you …

Our best possible price today is …

Please find enclosed some information about …

We can offer a large selection of …

We are able to quote you very advantageous terms.

Terms

Attached is the estimate you requested. Please note that these numbers are subject to change.

This offer is subject to …

Our usual terms are …

For orders of more than …, we can offer you …

If you can increase your order to … we can offer you …

Replying negatively to product inquiry

We are currently out of x and we do not know when we will be getting more.

We don't have a dealer in your country but *company* may be able to help you.

We regret that we no longer stock this item.

Unfortunately we are currently not in a position to supply you with the goods you requested.

We regret to inform you that due to … we will not be able to supply you with …

Apologizing for being unable to give requested information

I am sorry not to be able to give you the information requested.

We are sorry to tell you that …

We are sorry not to be able to give you a definite reply regarding the above matter.

Asking for confirmation

Please email / fax confirmation of the above order.

We look forward to receiving confirmation of this order.

Giving contact details

For any technical questions please contact …

If you would like further information on … or if you would like to be added to our mailing list … please mail me at the above address.

17.3   Making and canceling orders

Making an order

We have pleasure in sending you our order for …

Attached please find our order for …

Please supply the following items:

We would like to place the following order:

Setting conditions for possible purchase

If the quality of your products matches our requirements …

If you can guarantee a delivery by 18 June / within a month …

The device must be dispatched by courier.

We are placing this order on condition that the goods reach us by March 10 at the latest.

Please ensure that the goods reach us by the end of this month.

We reserve the right to refuse the goods if received after March 31.

Asking about delivery

Could you let me know the shipping cost?

How long will delivery take?

When can we expect delivery?

Do you ship direct to …?

How can I track my order/shipment?

Please arrange for delivery by …

Stating your terms

As the goods are required urgently, we would be grateful for delivery by …

We must insist on delivery within the time stated and reserve the right to reject the goods should they be delivered later.

Please forward the consignment by *means of transport.*

Please arrange for delivery by *means of transport.*

Giving details for the invoice

Please invoice us at:

Our full address is:

Our VAT number is:

Changing / canceling an order

Could you possibly make the following changes to our order:

We have to cancel the above mentioned order.

I regret to have to ask you to ...

We regret that we have to cancel the order contained in our letter of ...

This is due to circumstances beyond our control.

17.4   Accepting an order, giving details

Giving confirmation

We are pleased to confirm the above order.

We confirm the following terms:

Stating terms

Our usual terms are ...

For orders over 600 we can give you a special discount of 10 %.

This offer is firm for 10 days.

The offer is subject to ...

This particular order is not subject to the usual discounts.

Giving shipment details

Shipment is made by express post or registered air.

We will confirm any charges with you prior to shipping.

We can guarantee delivery by the time stated.

We assure you that the goods will be insured at the best possible rates.

Checking details

Could you confirm a couple of things in your order for me so I can get it posted to you as soon as possible.

It's not completely clear from your order whether ...

Delivery details / costs

The shipping options to Germany are as follows: surface mail, 4–6 weeks, $ 3.97: or airmail, 4–7 days, $ 10.80.

If we send the goods via / through air mail it will cost about $ 17.00 USD.

Delivery will be made on ...

We will invoice you upon shipment.

Stating when shipment will be / has been made

The goods were forwarded today by air.

We are pleased to inform you that your goods were shipped / dispatched today.

As requested, the shipping documents have been sent to …

Delivery will be made on March 10 / next Wednesday / by the end of the month.

The goods were forwarded today by *means of transport*.

The goods will be sent tomorrow by courier.

Stating next step

You will be emailed within 24 hrs of delivery to confirm that the order has been received.

After receipt of your payment we will immediately email confirmation of your order.

17.5   Contracts

Enclosing

Please find enclosed the offer and an unsigned copy of the standard agreement. This is for your perusal only.

Herein enclosed you will find two copies of the contract.

Explaining content

Please note that the annual fee includes maintenance and service.

The start date of the contract is *date*. The contract will be automatically renewed unless you notify us in writing one month in advance.

A contract can be transferred and registered for another company. In such case, please notify us of the name of the new company and the address.

Please let me know if you see the need for any additions or deletions.

Telling client the next step

Please return one of the copies, duly signed by you.

Please sign one copy of the pdf digitally.

Please sign another copy by hand and post it to us by recommended delivery to the following address:

17.6   Chasing orders

Explaining current status

With reference to our order #456 dated 10 March, we wish to inform you that as of today we have not received the goods.

We have not yet received the order for *product* (offer No. 123 dated 6 July, sent by email / post / courier).

The goods we ordered on 20 October have not yet been received.

Our order No. 123 for *product* which should have been delivered on *date* and is now considerably overdue.

Explaining consequences

Unless the goods can be dispatched immediately ...

If the order is not sent without further delay, we shall be obliged to cancel the order.

Asking for expected delivery time

As the *product* is urgently needed I would be glad if you would dispatch it without further delay.

Please let us know by the end of today when we can expect delivery.

17.7   Replying to chase of order

Apologizing

We very much regret that we have been unable to complete your order by the ...

Unfortunately we are unable to deliver more than ...

We are very sorry about this delay in the delivery of your order.

We regret that *product* ...

   ... is no longer available.

   ... will not be available before the ...

   ... cannot be delivered within the time specified.

Explaining reason for delay

Deliveries have been held up due to ...

   ... a shortage of raw materials.

   ... the breakdown of a machine.

... a strike.

... staff absence owing to the flu epidemic.

... circumstances beyond our control.

Unfortunately the main operating machine has broken down.

There is a heavy staff absence due to ...

The execution of your order has been delayed by circumstances beyond our control.

Explaining when order will be delivered

We regret that owing to ... we are still unable to deliver your order No. 123.

We are making every effort to execute your order as soon as possible.

We hope to deliver part of your order on ...

The balance will follow the first week of next month.

Apologizing again

We are very sorry that the delay has caused you so much inconvenience.

17.8   Payment details

Enclosing invoice

We have pleasure in sending you herewith statement of your account.

Attached is our invoice amounting to ...

We are pleased to attach our monthly statement.

Please find attached our invoice for the supply of ...

We attach our invoice in duplicate / triplicate.

We are attaching an invoice for the total amount of the goods.

Means of payment

I would be glad if you would remit the amount of the invoice to the ...

Please could you credit our account with the amount.

A personal check or money order will be fine if you order by mail.

Visa or Mastercard can be used for orders by e-mail, phone, or fax.

At this time, credit cards are the only method of payment accepted.

All the charges of the Wire Transfer must be borne by you.

Terms of payment

Payment terms are net thirty (30) days, unless otherwise provided.

Please make all payments to:

If you pay by credit transfer, payments should be made as follows:

Please send remittance advice to:

Payment for all purchases should be made by the end of the month following the date of the invoice (full payment details are attached).

Please see to the immediate transfer of the amount shown in the attached invoice.

All prices are subject to change without notice.

17.9   Chasing payment

Reminding

I am writing to remind you that we have still not received payment for …

May we remind you that we are still awaiting your reply to our message dated … re …

I wish to remind you of our letter dated 3 June in which we asked you to …

According to our records the following invoice has not been paid:

Our invoice # 234 dated 10 March is overdue. We would be grateful if you could settle it within the next ten days.

Our accounts status shows that payment for one of your packages is pending and needs to be cleared at the earliest. The details for the same are given below. You may also check these details by logging into your account at http://order.zzxx.com/account.

Please give the matter your immediate attention.

We enclose a copy of your quarterly statement dated 30 June which shows an unpaid balance of $ 1,999.

Re-contacting about the same issue

We are surprised that we have still not received a reply to our email dated 30 May in which we ask you to kindly settle our invoice No. 123 which was due three months ago.

I would like to inform you that our invoice dated 13 November is still unpaid. We enclose a copy of the statement of account.

We regret to have to remind you once again about ...

Sorry to bother you again. Sharon is our contact for the invoices, but is she also the person who we should be sending the contracts to?

Consequence of not receiving payment

Given that it was agreed that all invoices should be settled within 30 working days, we must insist on an immediate settlement.

Should we not receive payment within 10 working days, we will be forced to take legal action.

Unless you settle the account by the end of this month, we shall be forced to take legal proceedings.

If we do not receive payment by the end of this week, we will have to ...

> ... charge you interest.

> ... cancel your licence.

> ... take legal action.

Concluding

Could you please let me know as soon as possible when we can expect payment.

Could you please get back to me either later today or tomorrow morning.

Thanks for your help.

Please let me know if there's a problem with this so I can ...

Asking for acknowledgement

Kindly acknowledge receipt.

Please could you acknowledge receipt of this email.

17.10 Replying to a request for payment

Apologizing for delay in payment

Apologies for the delay in replying. Please find enclosed ... as requested.

I would like to apologize for the delay in responding to ...

I regret to inform you that we are unable to settle our account by the agreed date.

Stating that payment is being made or will be made

We acknowledge receipt of your invoice re ...

We have instructed our bank to make a credit transfer of € 10,000 in settlement of your invoice.

We have pleasure in sending you herewith our check for … in payment of your invoice No. 13 dated 10 March.

We have arranged payment of … through *name of bank* in settlement of …

We will be processing your invoice within the next ten days.

Stating that payment has already been made

We are pleased to inform you that your invoice was paid this morning.

Your invoice has in fact already been paid. I will fax you the bank details later this morning.

Your invoice No 18 dated 17 September was settled on 17 October by bank transfer (see attached transfer note).

Inaccuracies in the invoice or price

On checking your invoice dated 10 March we …

The item should read xxx not yyy.

No doubt it is an oversight on your part but …

We would like to point out that …

Some mistake must have been made.

The reason that your invoice has not been paid is that

   … it has been dated last year rather than this year.

   … VAT has been calculated at 20 % rather than 21 %.

   … the amount does not correspond to the amount on the order.

   … it has the wrong department in the address.

If you could please reissue the invoice, we will then ensure prompt payment.

17.11   Reporting problems with goods received

Announcing the problem

It has been brought to our attention that …

On checking the goods we found that they had been damaged during shipment.

You appear to have sent us the wrong order.

The following items are missing:

The goods you've just sent us don't seem to be up to your usual standards.

We are sorry to have to tell you that the goods arrived in a very poor condition.

The goods are not according to the samples

The service you are providing is not in accordance with the conditions of the order.

To our surprise we found that ...

Generic details

Unfortunately, there's a problem with ...

We haven't received ...

There seem(s) to be ...

The *product* doesn't work.

The quality of the work is below standard.

Requesting action

Could you deliver the missing article tomorrow morning?

Please rearrange shipment by air since we need the goods very urgently.

We must ask you to ...

   ... replace the damaged goods.

   ... credit us with the value of the damaged / returned goods.

Adding details that highlight the negative situation

It's not the first time we've had this problem.

This is the second time this has happened.

Six weeks ago, you assured us that ...

We must point out that this delay is causing us serious problems.

Outlining possible consequences

If the problem is not resolved by the end of this week, the consequences could be very serious.

We'll have to ...

   ... reconsider our position.

... contact other suppliers.

... renegotiate the contract.

We reserve the right to ... if we do not receive the goods from you by the end of this month.

17.12   Replying to customer complaints

Apologizing

We were sorry to hear that you were not satisfied with your last order.

We are looking into the matter and hope to rectify the situation as soon as possible.

Please accept our apologies for any inconvenience this may have caused you.

We apologize for having charged you the wrong amount.

I am very sorry, we should have called you immediately.

We are very sorry indeed to hear that ...

We are sorry for the inconvenience this may have caused you.

Please accept our apologies.

Giving an update

Since our last email to you on 19 August, we have ascertained that ...

Since receipt of your fax dated 6 March, we have been trying to find out more about the ...

In reply to our inquiry on your behalf we have been told by *person* that ...

According to information received from *person* it appears that ...

Please note that ...

Reassuring

I assure you that you will receive the goods by the end of the week.

We are happy to send replacements immediately.

We assure you that we are doing our best to set things right as soon as possible.

Concluding and apologizing again

We hope this inconvenience will not affect our good business relations.

We hope this delay will not cause you too much trouble.

Countering the complaint

We are very surprised at your complaint.

We have thoroughly examined the items you sent back to us and can find no fault with them.

We are still not clear exactly what seems to be the problem.

This is the first time we have received such a complaint.

According to our records, this was in fact the amount agreed.

Offering further clarifications

If you need any further details do not hesitate to contact me.

Should you have any questions please let us know.

Please do not hesitate to contact us should you need any further information.

Please contact me if you need any further clarifications.

Adrian Wallwork

I am the author of over 30 books aimed at helping non-native English speakers to communicate more effectively in English. I have published 13 books with Springer Science and Business Media (the publisher of this book), three Business English course books with Oxford University Press, and also other books for Cambridge University Press, Scholastic, and the BBC.

I teach Business English at several IT companies in Pisa (Italy). I also teach PhD students from around the world how to write and present their work in English. My company, English for Academics, offers an editing service.

Contacts and Editing Service

Contact me at: adrian.wallwork@gmail.com

Link up with me at:

www.linkedin.com/pub/dir/Adrian/Wallwork⬚

Learn more about my services at:

e4ac.com

A. Wallwork, *Email and Commercial Correspondence,*
Guides to Professional English, DOI 10.1007/978-1-4939-0635-2,
© Springer Science+Business Media New York 2014

Index

This index is by section number, not by page number. Numbers in bold refer to whole chapters. Numbers not in bold refer to sections within a chapter.

A. Wallwork, *Email and Commercial Correspondence,*
Guides to Professional English, DOI 10.1007/978-1-4939-0635-2,
© Springer Science+Business Media New York 2014

164